T0211672

Lecture Notes in Mathematics

Volume 2303

CEMPI CENTRE EUROPÉEN
POUR LES MATHÉMATIQUES, LA PHYSIQUE ET
LEURS INTERACTIONS

CEMPI is a joint project for research, training and technology transfer of the Laboratoire de Mathématiques Paul Painlevé and the Laboratoire de Physique des Lasers, Atomes et Molécules (PhLAM) of the Université Lille 1 and the CNRS. It was created as a "Laboratoire d'Excellence" in the framework of the "Programme d'Investissements d'Avenir" of the French government in February 2012.

Research at CEMPI covers a wide spectrum of knowledge from pure and applied mathematics to experimental and applied physics. CEMPI organizes every year the Painlevé-CEMPI-PhLAM Thematic Semester that brings together leading scholars from around the world for a series of conferences, workshops and post-graduate courses on CEMPI's main research topics.

For more information, see CEMPI's homepage http://math.univ-lille1.fr/~cempi/.

CEMPI Scientific Coordinator and CEMPI Series Editor
Stephan DE BIÈVRE (Université Lille 1)

CEMPI Series Editorial Board
Prof. Fedor A. BOGOMOLOV (New York University)
Dr. Jean-Claude GARREAU (Université Lille 1 and CNRS)
Prof. Alex LUBOTZKY (Hebrew University)
Prof. Matthias NEUFANG (Carleton University and Université Lille 1)
Prof. Benoît PERTHAME (Université Pierre et Marie Curie)
Prof. Herbert SPOHN (Technische Universität München)

Najib Idrissi

Real Homotopy of Configuration Spaces

Peccot Lecture, Collège de France, March & May 2020

 Springer

Najib Idrissi
CNRS, IMJ-PRG
Université Paris Cité and Sorbonne
Université
Paris, France

This work was supported by IdEx University of Paris (ANR-18-IDEX-0001).

ISSN 0075-8434   ISSN 1617-9692   (electronic)
Lecture Notes in Mathematics
ISBN 978-3-031-04427-4   ISBN 978-3-031-04428-1   (eBook)
https://doi.org/10.1007/978-3-031-04428-1

Mathematics Subject Classification: 55R80, 55P62, 18M75, 57R19

This Springer imprint is published by the registered company Springer Nature Switzerland AG
The registered company address is: Gewerbestrasse 11, 6330 Cham, Switzerland

À Ayman et Maman.
Ne craignez pas la tristesse car elle est la trace éclatante que quelque chose de beau a existé. *(Baptiste Beaulieu)*

Foreword

This book is based on the Peccot Lectures given by the author, Najib Idrissi, at the Collège de France in March 2020. The purpose is to give an account of recent results on the homotopy of configuration spaces of manifolds obtained using methods of real homotopy theory. These results were for one part obtained by the author during his thesis in Lille, for another part obtained by Ricardo Campos and Thomas Willwacher in related research on graph complexes. The advances led to subsequent collaborations, which are also reviewed in this book. The obtained results now provide a whole set of methods and statements that give a comprehensive understanding of configuration spaces of manifolds in the setting of real homotopy theory. I am delighted by this opportunity to write this foreword in order to introduce this work.

The configuration spaces of a topological space M explicitly consist of the spaces $\mathsf{Conf}(M, r) = \{(x_1, \ldots, x_r) \in M^{\times r} | x_i \neq x_j (\forall i \neq j)\}$, whose elements are the ordered collections $(x_1, \ldots, x_r)$ of pairwise distinct points $x_i \in M$, $i = 1, \ldots, r$. The study of configuration spaces, which lies in the scope of algebraic topology, turned out to be very difficult in spite of the simplicity of this definition.

This difficulty can be explained by observing that the configurations spaces $\mathsf{Conf}(M, r)$ do not form homotopy invariants in general, at least without making particular assumptions on the topology of the space M. Briefly recall that continuous maps $f, g : M \to N$ are homotopic $f \sim g$ if we have a continuous family of maps $h_t : M \to N$, $t \in [0, 1]$, so that $h_0 = f$ and $h_1 = g$, and that topological spaces M and N are homotopy equivalent $M \sim N$ if they can be connected by maps which are inverse to each other up to this homotopy relation. The most usual invariants of algebraic topology, like the classical singular homology theory of spaces, are homotopy invariants in the sense that these invariants are the same for homotopic maps and homotopy equivalent spaces.

For the configuration spaces, the question is whether homotopy equivalent spaces $M \sim N$ have homotopy equivalent configuration spaces $\mathsf{Conf}(M, r) \sim \mathsf{Conf}(N, r)$. The answer to this question is negative for general spaces, as can be seen on easy examples (e.g., Euclidean spaces), and the idea is to focus on the case of homotopy equivalent compact manifolds. But still Paolo Salvatore and Ricardo

Longoni have produced an example of homotopy equivalent 3-manifolds (lens spaces) whose configuration spaces are not homotopy equivalent. This example consists of non simply connected manifolds. The homotopy invariance of configuration spaces remains therefore open for simply connected compact manifolds. In any case, this discussion stresses the difficulty of understanding the homotopy of configuration spaces through methods of algebraic topology, since we have to take care of fine homotopical structures attached to manifolds; therefore, we can not only rely on basic homotopy invariants to address this question.

The works explained in this book provide solutions to this problem in the context of real homotopy theory, which is an approximation of the homotopy theory of spaces. This approximation encompasses the information that can be captured by a model, which is formed in the category of differential graded commutative algebras. (I explain this concept shortly.)

This notion of approximation is easier to explain in the setting of rational homotopy theory first. The general idea is to adapt the localization methods of algebra to topological spaces. The rational homotopy approximation captures the information governed by divisible homotopy invariants, for instance, by considering the singular homology with rational coefficients (the rational homology) instead of the singular homology with integral coefficients. In fact, in the simply connected case, spaces are rationally homotopy equivalent $X \sim_{\mathbb{Q}} Y$ if they can be linked by a zigzag of maps that induce an isomorphism in rational homology. In this case, we also say that the spaces X and Y have the same rational homotopy type. There is also a universal space $X_{\mathbb{Q}}$, associated to any space X, which captures this rational information and which forms a topological counterpart of the classical rationalization functor $A_{\mathbb{Q}} = A \otimes_{\mathbb{Z}} \mathbb{Q}$ in the realm of algebra. The real homotopy type is similarly encoded by a realification functor $X_{\mathbb{R}}$, which forms a topological counterpart of the scalar extension $A_{\mathbb{R}} = A \otimes_{\mathbb{Z}} \mathbb{R}$ of algebra.

The idea of a model, given by the Sullivan rational homotopy theory, is that we can reconstruct the rationalization of a space $X_{\mathbb{Q}}$ from an associated differential graded commutative $\mathbb{Q}$-algebra $A(X) = A^*$, a commutative algebra (in the graded sense) equipped with a differential $d : A^* \rightarrow A^{*+1}$ which acts as a derivation (in the graded sense too). There is an analogous statement for the realification, considering differential graded commutative $\mathbb{R}$-algebras instead of differential graded commutative $\mathbb{Q}$-algebras. For a manifold M, one can precisely prove that the de Rham cochain algebra of differential forms $\Omega^*_{dR}(M)$ gives a model of the real homotopy type of the space X in this sense. But every differential graded commutative $\mathbb{R}$-algebra A that we can link to $\Omega^*_{dR}(M)$ by a zigzag of quasi-isomorphisms (morphisms of differential graded commutative $\mathbb{R}$-algebras that induce an isomorphism at the cohomology level) gives a model of the real homotopy type too, because the quasi-isomorphisms, on the differential graded commutative algebra side, reflect real equivalences on the topological side. The general idea is to take advantage of the computability features of algebra in order to associate to M a small model A, which retains only the essential information of the differential graded algebra $\Omega^*_{dR}(M)$, and is therefore enough to determine the real homotopy type with minimal information. This is precisely what has been achieved in the case of configuration spaces. I would like to give a flavor of these results, which give the core of this book.

The starting point of the construction is a statement obtained by Maxim Kontsevich in the case of the configuration spaces of Euclidean spaces $\mathsf{Conf}(\mathbb{R}^n, r)$ and which asserts that, in this case, a Sullivan model is provided by the cohomology of the space $A^* = H^*(\mathsf{Conf}(\mathbb{R}^n, r), \mathbb{R})$, regarded as a differential graded commutative algebra equipped with a trivial differential $d = 0$. In the language of the rational homotopy theory, this result asserts that the configuration spaces of the Euclidean spaces are formal. The structure of the cohomology algebra $H^*(\mathsf{Conf}(\mathbb{R}^n, r), \mathbb{R})$ is itself well known, as the exterior algebra generated by classes ω_{ij} of degree $n - 1$, defined for pairs $\{i, j\} \subset \{1, \ldots, r\}$ with $\omega_{ij} = (-1)^n \omega_{ji}$, and moded out by the ideal generated by relations $\omega_{ij}\omega_{jk} + \omega_{jk}\omega_{ki} + \omega_{ki}\omega_{ij} \equiv 0$, first established by Arnold in the 2-dimensional case.

The result of Najib Idrissi, in the case of a closed compact smooth simply connected n-manifold M, precisely asserts that a model $G_A(r)$ of the configuration space $\mathsf{Conf}(M, r)$ is given by an extension of this exterior algebra $\wedge(\omega_{ij})/(\text{Arnold})$ by an r-fold tensor product of a Poincaré duality model $A = A(M)$ of the manifold M together with a differential yielded by a rule $d(\omega_{ij}) = p_{ij}^*(\Delta_M)$, where $\Delta_M \in A \otimes A$ represents the diagonal class of the manifold in this Poincaré duality model $A = A(M)$ and $p_{ij}^* : A^{\otimes 2} \to A^{\otimes r}$ puts this tensor at position (i, j) in the tensor product $A^{\otimes r}$.

This Poincaré duality model $A = A(M)$ is just a differential graded algebra equipped with a perfect pairing $\langle -, - \rangle : A^* \otimes A^{n-*} \to \mathbb{R}$, compatible with the algebra structure, and which reflects the Poincaré duality pairing $\langle -, - \rangle : H^*(M, \mathbb{R}) \otimes H^{n-*}(M, \mathbb{R}) \to \mathbb{R}$ at the cohomology level. Thus, the existence of this Poincaré duality model is only what we need to retain from the structure of the manifold M in order to reconstruct a model of the real homotopy type of the configuration space $\mathsf{Conf}(M, r)$. In fact, as this Poincaré duality model is unique (up to quasi-isomorphism), we also conclude from the construction of the model $G_A(r)$ that the configuration space $\mathsf{Conf}(M, r)$ is a real homotopy invariant of the manifold M (in the simply connected case). Hence, we obtain an answer to the homotopy invariance question raised in the introduction of this foreword in the setting of real homotopy theory.

The proof of the validity of this model of configuration spaces $G_A(r)$ goes through an intermediate model, which was defined by Campos–Willwacher in their work, and which generalizes graph complexes used by Kontsevich in his proof of the formality of configuration spaces. These graph complexes govern differential forms defined in a semi-algebraic variant of the usual de Rham complex of differential forms. In the case of the configuration spaces of manifolds, the definition of the graph complex involves a partition function on configurations Z_ϕ, which depends on the choice of a propagator function for configurations of two points ϕ, and one of the cruxes of the proof of the validity of the small model $G_A(r)$ is a proof that we get equivalent results in homotopy if we replace this propagator ϕ by a trivial one ϵ.

The book also explains an extension of these results in the context of manifolds with boundary. (Then, to get a counterpart of the model $G_A(r)$, we have to take a model that reflects the Lefschetz duality instead of the Poincaré duality.)

The Kontsevich formality result was obtained in the context of the theory of operads. The configuration spaces of (framed) n-manifolds inherit the structure of a module over the operad of little n-disks (roughly, a composition structure, which asserts that a configuration of points can be composed with certain configurations of disks in order to return a new configuration of points in the manifold). This structure is used in the definition of the factorization homology (a multiplicative analogue of the classical singular homology of spaces) and the book also explains the applications of the model of configuration spaces to the computation of factorization homology (in the real homotopy setting).

The book, after a basic introduction to the subject of configuration spaces, includes a comprehensive review of the general definitions, constructions, and results of homotopy theory that form the background of this study: rational homotopy theory, applications of differential forms, and theory of semi-algebraic forms on semi-algebraic sets. The applications of graph complexes are also explained with full details, as well as the definition of Poincaré duality and of Lefschetz duality models for manifolds, and the definition and applications of operad structures for the outlook chapter, about the applications to the factorization homology theory. This book is therefore an invaluable effort to make these significant advances on the homotopy theory of configuration spaces accessible to a wide audience of students and researchers. I hope very much that the reader will enjoy it as I did.

Lille, France   Benoit Fresse
February 2022

Preface

The goal of this volume is to present recent developments regarding the real homotopy type of configuration spaces of manifolds. Given a space M (usually a manifold) and an integer $r \geq 0$, the rth configuration space $\mathsf{Conf}_M(r)$ is the space of r ordered pairwise distinct points in M. These are classical objects that already implicitly appear in the work of Hurwitz [Hur91] (1891) and have been studied in detail since at least the 1960s [FN62a]. These spaces have appeared since then in numerous contexts: one can mention braid groups [FN62b], iterated loop spaces [BV68; May72], moduli spaces of curves [Get95], Gelfand–Fuks cohomology [CT78], Goodwillie–Weiss manifold calculus [BW18; Tur13], factorization homology [AF15; AFT17; Fra13; Lur09b], motion planning [Far03], and robotics [Far18; Ghr01].

Despite their apparent simplicity, configuration spaces remain intriguing objects that are ubiquitous in topology and geometry. mentioned above mainly use the homotopy type of configuration spaces. In other words, we only consider configuration spaces up to continuous deformations. A basic question is homotopy invariance, which has remained open for decades: if two spaces M and N can be continuously deformed into one another, then can the same be done for their configuration spaces $\mathsf{Conf}_M(r)$ and $\mathsf{Conf}_N(r)$? This is clearly wrong in general, but there is evidence that this is true for closed manifolds thanks to work of Bödigheimer et al. [BCT89], Bendersky and Gitler [BG91], Levitt [Lev95], Aouina and Klein [AK04], Kriz [Kri94], Totaro [Tot96], and Lambrechts and Stanley [LS04; LS08a]. A counterexample due to Longoni and Salvatore [LS05] forces us to restrict to simply connected manifolds.

In this volume, we plan to explain why this conjecture holds over the reals, that is, if we restrict our focus to algebro-topological invariants defined over the field $\mathbb{R}$. We will also explain how a generalization of this conjecture also holds for manifolds with boundary. This is achieved by proving a conjecture of Lambrechts and Stanley [LS08a]. They built a certain commutative differential-graded algebra (CDGA) in the closed case and they conjectured that it is a model of $\mathsf{Conf}_M(r)$ (in the sense of the real homotopy theory of Sullivan [Sul77]). This CDGA had previously appeared in various forms in the works of several authors, in particular Cohen and Taylor

[CT78], Kriz [Kri94], Totaro [Tot96], Berceanu et al. [BMP05], and Félix and Thomas [FT04]. One of the main goals of the volume is to show that this CDGA is a model in the closed case, and moreover that a generalization of the model exists in the case of manifolds with boundary. Since this CDGA only depends on the real homotopy type of M, this proves that real homotopy invariance holds. The fact that we have full models also enables one to make concrete computations about configuration spaces, for example, their cohomology or their homotopy groups.

The basis for the proof of this result is the proof of Kontsevich [Kon99] of the formality of the little disks operads. The proof involves many different ingredients:

   (i) Rational/real homotopy theory of Sullivan models [Sul77]
  (ii) The (Axelrod–Singer–)Fulton–MacPherson compactifications of configuration spaces [AS94; FM94; Sin04]
 (iii) The theory of piecewise semi-algebraic forms on semi-algebraic sets [HLTV11]
  (iv) The mathematical physics' notion of a propagator
   (v) Kontsevich integrals, a version of which originally appeared in the study of knot invariants [BC98]
  (vi) Combinatorial arguments on graphs complexes

We plan to introduce the theory behind all these ingredients in an accessible way, and explain how they all fit together to form the proof. We will also explain how a key point of the proof is a simple degree-counting argument that works for manifolds that are at least four-dimensional. This argument can be philosophically explained as the fact that the partition function of the Poisson-σ model on these manifolds is trivial. connected closed manifolds are spheres in low dimension (thanks in particular to the Poincaré conjecture), this is not true for manifolds with boundary. in dimension ≥ 4 for such manifolds.

In addition to the ingredients mentioned above, the theory of operads plays an important role. The notion of an operad was introduced around five decades ago to study iterated loop spaces [BV68; May72] and has seen, since then, a large number of uses. Briefly, an operad is an algebraic object which encodes a category of algebras (e.g., the category of associative algebras, or the category of commutative algebras). There is an intimate relationship between configuration spaces of manifolds and certain operads called the little disks operads. The salient point of this relationship is that the little n-disks operad acts (up to homotopy) on the configuration spaces of a framed n-manifold.

The results we plan to explain are all compatible, in an appropriate way, with the action of the little disks operads on configuration spaces. This compatibility has important consequences. In particular, it allows one to compute factorization homology, a homology theory for manifolds that classifies all fully extended topological quantum field theories according to a result of Lurie. (Note that dually, it would also enable one to compute embedding spaces between manifolds using Goodwillie–Weiss calculus, as was, e.g., recently done by Fresse–Turchin–Willwacher [FTW17] for Euclidean spaces using Kontsevich formality.) We also plan to explain how a result of Knudsen [Knu18b] about factorization homology with coefficients in higher enveloping algebras of Lie algebras can be recovered using the Lambrechts–Stanley model.

This volume is intended for graduate students and researchers in algebra, topology, or geometry. We try not to assume too much of knowledge from the reader, and whenever possible, we give pointers to the existing literature for background or further developments. Some background on algebraic topology remains required, and some knowledge of (differential) geometry will help understand the finer points of semi-algebraic theory.

Unless otherwise indicated, the results presented in these notes come from the two papers [CW16; CILW18; Idr19]. Few new results are contained in these notes, although some new illustrative examples are worked out.

Paris, France   Najib Idrissi
March 2021

Acknowledgments

The material in this volume is based on the Peccot Lecture (*Cours Peccot*) given by the author at the Collège de France in March and May 2020. The videos of the lecture are available at https://www.college-de-france.fr/site/cours-peccot/ guestlecturer-2019-2020__1.htm. I would like to express my gratitude to the Collège de France and the Fondation Claude-Antoine Peccot for giving me the unique opportunity of presenting my work during these four lectures. I thank Claire Voisin for her interest and insightful comments. I also thank the administrative and technical staff of the Collège de France for handling the effects of the COVID-19 pandemic as best as could be.

Many of the results presented in these notes were obtained as part of my PhD thesis, and I thank my advisor, Benoit Fresse, for his help and guidance. Some of the results were obtained jointly with other researchers: Ricardo Campos, Julien Ducoulombier, Pascal Lambrechts, and Thomas Willwacher. It has always been a pleasure working with them and I thank them for that. Finally, I thank Muriel Livernet for her support and for many helpful comments and suggestions.

I would like to thank Oscar Randal-Williams and Pedro Tamaroff for helpful remarks, and Adrien Brochier for helpful discussions.

The work referred to in these notes was carried out at the Laboratoire Paul Painlevé at Université de Lille, the Department of Mathematics at ETH Zurich, as well as at Institut de Mathématiques de Jussieu–Paris Rive Gauche and Université de Paris. I thank these institutions, as well as my former and current colleagues who have always given me a warm welcome.

Je souhaiterais remercier toutes les personnes qui m'ont soutenu durant la période difficile que ma famille et moi-même avons traversée durant la rédaction de ces notes.

This research was supported by the LabEx CEMPI (ANR-11-LABX-0007-01), the ANR project HOGT (ANR-11-BS01-002), the ERC project GRAPHCPX (StG 678156), and contributes to the IdEx University of Paris (ANR-18-IDEX-0001).

Contents

Chapter 1
Overview of the Volume

In this chapter, we give a broad overview of the rest of this volume and we explain the key concepts. In the later chapters, we study these concepts in more detail. We give references to these more detailed explanations throughout this overview.

First of all, we briefly introduce the main object of study in this volume, configuration spaces. We also introduce their history and their applications. We then explain the context of the homotopy invariance conjecture, which states that two homotopy equivalent simply connected closed manifolds have homotopy equivalent configuration spaces. We then explain the main results of the later chapters, which settles this conjecture over the field $\mathbb{R}$. We conclude by a brief explanation of the relationship between configuration spaces and operads, an algebraic device that encodes categories of algebras. Each of the four sections of this overview corresponds to one of the later chapters of the volume.

1.1 Configuration Spaces of Manifolds

Let M be a fixed topological space and $r \geq 0$ an integer. The rth (ordered) configuration space of M, denoted by $\mathsf{Conf}_M(r)$, consists in the ordered collections of r pairwise distinct points in M. See Fig. 1.1 for an illustration.

Configuration spaces are classical objects in algebraic topology, as we will see in Chap. 2. They have been studied since the 1960's [FN62a; FN62b], and they already appeared implicitly in the work of Hurwitz [Hur91] in 1891 (see [Mag74]) as well as in the work of Artin [Art47]. Although they initially appeared in the study of braid groups (see Example 2.6), configuration spaces have proved useful in numerous settings. As an example of particular interest to us, they can be used to define invariants of manifolds, i.e., locally Euclidean spaces.

Despite the apparent simplicity of their definition, configuration spaces are intriguing objects. One of the main questions in the study of configuration spaces is

© The Author(s), under exclusive license to Springer Nature Switzerland AG 2022
N. Idrissi, *Real Homotopy of Configuration Spaces*, Lecture Notes
in Mathematics 2303, https://doi.org/10.1007/978-3-031-04428-1_1

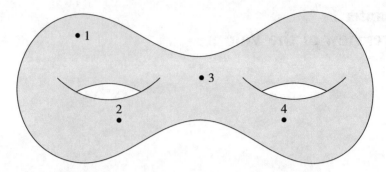

Fig. 1.1 An element of $\mathrm{Conf}_{\Sigma_2}(4)$, where Σ_2 is the oriented surface of genus 2

to know the extent of how the structure of M determines the structure of $\mathrm{Conf}_M(r)$. It is clear that two homeomorphic (resp. diffeomorphic) manifolds have homeomorphic (resp. diffeomorphic) configuration spaces. However, if two manifolds M and N are homotopy equivalent, then it is not clear that the configuration spaces of M and N are homotopy equivalent (see Sect. 2.2). Homotopy equivalences are not always injective and therefore may not even induce maps between configuration spaces.

If the question is stated in such a naive manner, then it is in fact clear that the answer is no. For example, the real line $\mathbb{R}$ is homotopy equivalent to the singleton $\{0\}$. However, the configuration space of two points in the real line $\mathrm{Conf}_{\mathbb{R}}(2)$, i.e., a plane with a diagonal removed, is not homotopy equivalent to the configuration space of two points in a singleton $\mathrm{Conf}_{\{0\}}(2)$, which is empty. There also exist counterexamples given by open manifolds of the same dimension (see Example 2.23). It was conjectured for years that, if this question was restricted to *closed* manifolds, i.e., compact manifolds without boundary, then the answer would be yes. However, after the discovery of a counterexample by Longoni and Salvatore [LS05], the conjecture was restricted even further to *simply connected* closed manifolds and remains open to this day. An analogous conjecture about configuration spaces of simply connected manifolds with boundary is also open.

Recall that a homotopy equivalence is always a weak homotopy equivalence, that is, it induces an isomorphism on homotopy groups. Two spaces are said to be weakly (homotopy) equivalent if they can be connected by a zigzag of weak homotopy equivalences, and the *homotopy type* of a space is its equivalence class under this relation. For good spaces (e.g., manifolds or CW complexes) this is equivalent to the existence of a direct homotopy equivalence (see Theorem 2.39). In this volume, we will focus on only a part of the homotopy types of the spaces involved. Namely, we are going to study the *rational* (or *real*) homotopy types of our spaces. The idea of rational homotopy theory—i.e., the study of rational homotopy types of spaces— goes back to the work of Serre [Ser53] and was developed by Quillen [Qui69], and Sullivan [Sul77]. In short, we say that two spaces have the same *rational* homotopy type if they can be connected by a zigzag of maps which induce isomorphisms on homotopy groups after they have been tensored with $\mathbb{Q}$ (that is, on the rational homotopy groups $\pi_*^{\mathbb{Q}}(-) = \pi_*(-) \otimes_{\mathbb{Z}} \mathbb{Q}$). We refer to Sect. 2.3 for the details.

Although one obtains less information on topological spaces in the framework of rational homotopy theory, there is an upside: rational homotopy types have the undeniable advantage of being *computable*. Indeed, the rational homotopy type of a (nice enough) topological space is completely determined by purely a algebraic piece of data called a model. A model of a space X is a commutative differential-graded algebra (CDGA) with rational coefficients which is quasi-isomorphic to the CDGA $\Omega_{PL}^*(X)$ of piecewise polynomial forms on X (see Definition 2.60). One can show that two spaces have the same rational homotopy type if and only if they admit quasi-isomorphic models (Corollary 2.66). A model of X can be used to perform numerous computations, e.g., the rational cohomology of X and its cup product, the rational homotopy groups of X and their Whitehead brackets, the Massey products on the cohomology, and so on.

In the framework of real homotopy theory, rational coefficients are replaced by real coefficients. Real homotopy types are defined in terms of models as explained in the previous paragraph (see Sect. 2.3 for the details). Real models contain slightly less information than rational models, because two spaces can have the same real homotopy type but different rational homotopy types (see Example 2.84). However, real models remain sufficient for most computations. One of our main goals, in this volume, will be to find models for the real homotopy types of configuration spaces of manifolds. Let us now turn our attention back to configuration spaces.

Manifolds are all built with the elementary "brick" given by the Euclidean spaces $\mathbb{R}^n$. It is thus natural to start with the study of configuration spaces of Euclidean spaces. The cohomology of the configuration spaces of $\mathbb{R}^n$ have been known since the work of Arnold [Arn69]. We recall the computation in Theorem 2.88. Briefly, the algebra $H^*(\mathsf{Conf}_{\mathbb{R}^n}(r))$ admits a presentation which reflects the fact that $\mathsf{Conf}_{\mathbb{R}^n}(r)$ is obtained from $(\mathbb{R}^n)^r$ by removing the diagonals $\Delta_{ij} := \{x \in (\mathbb{R}^n)^r \mid x_i = x_j\}$, whose intersections satisfy $\Delta_{ij} \cap \Delta_{jk} = \Delta_{jk} \cap \Delta_{ki} = \Delta_{ki} \cap \Delta_{ij}$. More precisely, $H^*(\mathsf{Conf}_{\mathbb{R}^n}(r))$ is generated by classes ω_{ij} (for $1 \leq i \neq j \leq r$) of degree $n - 1$ which satisfy the relations $\omega_{ij}^2 = 0$, $\omega_{ji} = (-1)^n \omega_{ij}$, and, for $i \neq j \neq k \neq i$, the three-term relations (commonly called "Arnold relations") $\omega_{ij}\omega_{jk} + \omega_{jk}\omega_{ki} + \omega_{ki}\omega_{ij} = 0$.

Arnold [Arn69] moreover proved that $H^*(\mathsf{Conf}_{\mathbb{R}^2}(r))$ is a model of $\mathsf{Conf}_{\mathbb{R}^2}(r)$ through an ad-hoc argument using complex analysis (see Sect. 2.4). This is an instance of a special situation called *formality* (a space X is called formal if $H^*(X)$ is a model of X). Arnold's proof, however, does not easily generalize to higher dimensions. However a different, more involved proof of the formality of the spaces $\mathsf{Conf}_{\mathbb{R}^n}(r)$ (for all $n \geq 2$ and all r) was found by Kontsevich [Kon99] (see also Lambrechts and Volić [LV14]). We thus get a simple, explicit description of the rational homotopy type of $\mathsf{Conf}_{\mathbb{R}^n}(r)$. The proof is outlined in Sect. 5.4.1 (it is in the last chapter as we are only going to need it for the operadic structure on configuration spaces, see the overview below). This more general proof will be used as a basis for the determination of models of configuration spaces of compact manifolds in the second and third chapters of this volume.

1.2   Configuration Spaces of Closed Manifolds

Let us now turn our attention to closed manifolds, i.e., compact manifolds without boundary. We also restrict ourselves to simply connected manifolds, i.e., ones in which closed loops can all be contracted continuously to a point. In Chap. 3, we will settle the homotopy invariance conjecture over $\mathbb{R}$ for this class of manifolds.

The model of $\mathsf{Conf}_{\mathbb{R}^n}(r)$ found at the end of the previous section was used as a foundation to build candidates for models of configuration spaces of simply connected closed manifolds (see Sect. 3.1). These candidates appeared in various shapes in the works of several authors [BG91; BMP05; Cor15; CT78; FT04; Kri94; LS04; LS08a; Tot96] (see also Remark 3.19). They are built in the following way. Let M be a simply connected closed manifold of dimension n. We can find a model A of M which satisfies Poincaré duality at the level of cochains thanks to a theorem of Lambrechts and Stanley [LS08b]. The candidate model of $\mathsf{Conf}_M(r)$, denoted $\mathsf{G}_A(r)$, is built from $H^*(\mathsf{Conf}_{\mathbb{R}^n}(r))$ by adding a tensor factor $A^{\otimes r}$ and by modding out by the relations $p_i^*(a)\omega_{ij} = p_j^*(a)\omega_{ij}$ (for $1 \leq i \neq j \leq r$ and $a \in A$). These relations reflect the fact that the two canonical projections $p_i, p_j : M^r \rightrightarrows M$ are equal when they are restricted to the diagonal Δ_{ij}.

Unlike $\mathsf{Conf}_{\mathbb{R}^n}(r)$, the spaces $\mathsf{Conf}_M(r)$ are rarely formal for closed manifolds M. The model $\mathsf{G}_A(r)$ must thus have a nonzero differential. This differential is the sum of the internal differential (induced by d_A) and the unique derivation which maps ω_{ij} to the class corresponding to the diagonal Δ_{ij}, also (abusing notation) denoted by $\Delta_{ij} \in A \otimes A$.

This model $\mathsf{G}_A(r)$ is conjectured to be a rational model of $\mathsf{Conf}_M(r)$, see [FOT08, Conjecture 9.9]. Since it only depends on A, a positive answer to this conjecture would lead to a positive answer for the rational homotopy invariance conjecture. The main result of Chap. 3, which comes from the article [Idr19], is the following.

Theorem 1.1 *Let M be a smooth simply connected closed manifold and let $r \geq 0$ be an integer. Let A be a Poincaré duality model of M. The CDGA $\mathsf{G}_A(r)$ is a real model of $\mathsf{Conf}_M(r)$.*

Corollary 1.2 *If two smooth simply connected closed manifolds have the same real homotopy type, then so do their configuration spaces.*

Similar results have been obtained simultaneously by Campos and Willwacher [CW16]. The proof, which takes up most of Chap. 3, is an adaptation and a generalization of the proof of Kontsevich [Kon99] mentioned above. One of the main ingredients of the proof is the theory of graph complexes, that we introduce in Sect. 3.4. Graph complexes as we use them were introduced by Kontsevich [Kon93] in the study of invariants of 3-manifolds. Since then, the interest for graph complexes has been growing considerably. We refer to Willwacher [Wil18] for a survey of the subject. For us, one of the main uses of graph complexes is that they provide a resolution (in the sense of homological algebra) of the Arnold and symmetry relations that define G_A. Indeed, as we will see, the CDGA of graphs is free as a graded commutative algebra. All the complexity is hidden in the differential, which can philosophically be interpreted as the fact that the Arnold and symmetry relations are satisfied up to homotopy in the graph complex.

Let us now quickly describe the graph complexes that we are going to use (see Sect. 3.4.2 for details). As a vector space, the graph complex is spanned by formal linear combinations of isomorphism classes of graphs. The graphs have two kind of vertices: numbered "external" vertices which represent static points in M, and indistinguishable "internal" vertices that represent mobile points in M. Each vertex is decorated by an element of a model of M. The differential consists in three summands: contraction of edges incident to internal vertices, application of the internal differential of the model to decorations, and cutting of edges and replacing them with diagonal classes, just like in $G_A(r)$. The model $G_A(r)$ is a quotient of this graph complex by the ideal generated by graphs containing internal vertices.

To show that graph complexes form a model of $\mathsf{Conf}_M(r)$, we use an integration procedure, on all the possible positions of the mobile points. To ensure that these integrals converge, it is necessary to compactify configuration spaces. The compactifications that we use, called the Fulton–MacPherson compactifications [AS94; FM94; Sin04], are obtained by adding a boundary to the open manifold $\mathsf{Conf}_M(r)$ (see Sect. 3.2). This boundary contains virtual configurations, where points are allowed to become infinitesimally close to one another. The integrals mentioned earlier are computed along the fibers of the canonical projections that forget the mobile points. When these projections are extended to the compactifications, they are not submersions any longer, but merely semi-algebraic (SA) fiber bundles. It is thus not possible to compute the integrals of usual de Rham forms along the fibers of these projections. To solve this problem, we use piecewise semi-algebraic (PA) forms. The study of such form was initiated by Kontsevich and Soibelman [KS00] and further developed by Hardt, Lambrechts, Turchin, and Volić [HLTV11]. We introduce briefly this theory in Sect. 3.3.

The full definition of the graph complex model above depends on similar integrals, indexed by graphs that only contain internal vertices. These integrals themselves depend, a priori, on the semi-algebraic structure of the manifold. Two homotopy equivalent manifolds may not always produce the same integrals. One of the key points of our proof is the fact that these integrals actually vanish when the manifold is simply connected and its dimension is at least 4. This allows us to establish the (real) homotopy invariance of our models in Sect. 3.5.

Remark 1.3 If the dimension of the manifold is less than 4, then the proof is completely different: the only possible manifolds are sphere, by the classification of surfaces and the Poincaré conjecture (see Perelman [Per02; Per03]).

1.3 Configuration Spaces of Manifolds with Boundary

In Chap. 4, which mainly relates the results of the article [CILW18] (joint with Campos, Lambrechts, and Willwacher), we generalize the previous results to manifolds with boundary. One of the main objectives of this endeavor is to provide a way to reconstitute the configuration spaces of a complicated manifold from the configuration spaces of simpler submanifolds. This idea is omnipresent in algebraic topology, and particularly in topological quantum field theory, which motivates the following. For example, one can obtain an oriented surface of genus g by gluing g cylinders on a sphere with $2g$ holes (see Fig. 1.2 for $g = 2$). This can classically be used to compute the cohomology of that surface, for example.

The classical case is when a manifold is of the form $M \cup_{N \times \mathbb{R}} M'$, obtained by gluing two submanifolds M, M' along a collar of their common boundary $N = \partial M = \partial M'$ (see Fig. 4.1). The collection of the configuration spaces $\mathsf{Conf}_{N \times \mathbb{R}} = \{\mathsf{Conf}_{N \times \mathbb{R}}(r)\}_{r \geq 0}$ is endowed with the structure of a monoid up to homotopy (see Fig. 4.2). Moreover, the collection Conf_M (resp. $\mathsf{Conf}_{M'}$) is endowed with the structure of a left (resp. right) module up to homotopy over $\mathsf{Conf}_{N \times \mathbb{R}}$ (see Fig. 4.3). The "derived tensor product" $\mathsf{Conf}_M \otimes^{\mathbb{L}}_{\mathsf{Conf}_{N \times \mathbb{R}}} \mathsf{Conf}_{M'}$ has the homotopy type of the collection $\mathsf{Conf}_{M \cup_{N \times \mathbb{R}} M'}$. This procedure allows us to compute homotopy types of configuration spaces in an inductive manner.

We start with some preliminaries on Poincaré–Lefschetz duality in rational homotopy theory, which is the generalization of Poincaré duality to manifolds

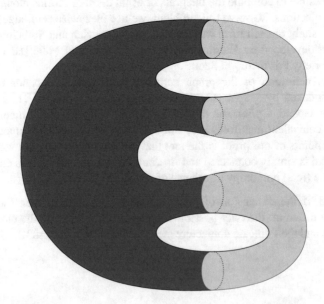

Fig. 1.2 The surface Σ_2 obtained by gluing a sphere with four holes (dark part) together with two cylinders (lighter part)

with boundary (see Sect. 4.2). Then, we introduce a model for configuration spaces which reflects these algebraic structure of monoid/module (Sect. 4.3). We will first describe compactifications of $\mathrm{Conf}_{N \times \mathbb{R}}$ and Conf_M which are similar to the Fulton–MacPherson compactifications. These compactifications form respectively a strict monoid and a strict module. Moreover, they enable us to compute integrals along fibers (although the projections remain semi-algebraic fiber bundles rather than submersions). Just like in Chap. 3, we will define models for these compactifications based on graph complexes. These models will be spanned by graphs whose vertices are decorated by a model of the pair $(M, \partial M)$. The monoid/module algebraic structure is given, on graph complex, by the operations which consist in partitioning a graph in two and in replacing cut edges by appropriate decorations.

These models still depend on integrals. We can show that the integrals used to define the model for $\mathrm{Conf}_{N \times \mathbb{R}}$ always vanish up to homotopy. This is consistent with previous results that show that the homotopy type of $\mathrm{Conf}_{N \times \mathbb{R}}$ only seems to depend on the homotopy type of N (see e.g., Raptis and Salvatore [RS18]). For the model of Conf_M, we need additional conditions on the dimension and connectivity of M, just like in Chap. 3. In summary, we obtain:

Theorem 1.4 ([CILW18]) *Let M be a compact manifold with boundary N. The graph complexes models of Chap. 4 are real models of $\mathrm{Conf}_{N \times \mathbb{R}}$ and Conf_M which are compatible with the algebraic structures of monoid/module.*

Corollary 1.5 *Let M be a compact manifold with boundary N. The real homotopy type of $\mathrm{Conf}_{N \times \mathbb{R}}$, as monoid in collections of topological spaces, only depends on the real homotopy type of N. If M is simply connected and $\dim M \geq 4$, then the real homotopy type of Conf_M, as a right module over $\mathrm{Conf}_{N \times \mathbb{R}}$ in collections of topological spaces, only depends on the real homotopy type of the pair (M, N).*

If we add the hypothesis that the boundary of M is simply connected, then in dimension ≤ 3 the only possible manifolds are disks by the Poincaré conjecture and the classification of surfaces. The conclusion of the corollary above thus becomes true in any dimension.

In the last part of the chapter (Sect. 4.4), we will introduce another model of Conf_M which is defined analogously to the model G_A of Chap. 3. Let M be a compact manifold with boundary. We start by showing that there exists a model of the pair $(M, \partial M)$ which satisfies Poincaré–Lefschetz duality at the level of cochains if M and ∂M are simply connected and $\dim M \geq 7$. We then define a CDGA G_P (where P is part of the Poincaré–Lefschetz duality model) analogously to the closed case. If $\partial M = \varnothing$, this CDGA is identical to the one of Chap. 3. If we assume that M and ∂M and simply connected, then we can use once again arguments of Lambrechts and Stanley [LS08a] and Cordova Bulens, Lambrechts, and Stanley [CLS18] to show that the cohomology of $\mathsf{G}_P(r)$ is isomorphic, as a graded representation of the symmetric group $\mathfrak{S}_r$, to the rational cohomology of $\mathrm{Conf}_M(r)$.

However, G_P is not always a model of Conf_M, as one can heuristically check on simple examples (e.g., $M = \mathbb{S}^1 \times [0, 1]$). We thus define in Sect. 4.4 a "perturbed" CDGA $\tilde{\mathsf{G}}_P$ which is isomorphic to G_P as a cochain complex but not as a CDGA.

Our terminology comes from the fact that, in the relations defining $\tilde{\mathsf{G}}_P$, the highest weight part precisely give the relations defining G_P. In many cases (e.g., if M is obtained by removing a point from a closed manifold) then $\tilde{\mathsf{G}}_P$ and G_P are actually equal. We can show, using the results of the first part of the chapter, that $\tilde{\mathsf{G}}_P$ is a model of Conf_M under appropriate conditions. We then obtain the following theorem:

Theorem 1.6 ([CILW18]) *Let M be a smooth simply connected compact manifold with boundary of dimension at least 5 and assume that M admits a Poincaré– Lefschetz duality model P. Then for all $r \geq 0$, the CDGA $\tilde{\mathsf{G}}_P(r)$ is a real model of* $\mathsf{Conf}_M(r)$. *The same result is true for* $\dim M \in \{4, 5, 6\}$ *with* $P = H^*(M)$.

Let us remark that the above results do not provide much information for configuration spaces of low-dimensional manifolds. The model of Theorem 1.4 depends on integrals that we do not always know how to compute, and Theorem 1.6 does not apply. In the article [CIW19] (joint with Campos and Willwacher), we provide a model for the configuration spaces of oriented surfaces by different methods.

1.4   Configuration Spaces and Operads

The proofs of the preceding results are all based on ideas coming from the theory of operads, which is the subject of Chap. 5. Operads were initially introduced at the end of the 1960s in algebraic topology to study iterated loop spaces [BV68; May72]. They have, since then, found numerous applications in several domains of mathematics (e.g., deformation quantization or homological algebra). We refer to Sect. 5.2 for more details. Briefly, an operad in an algebraic object that encodes a category of algebras, just like e.g., a group encodes a category of group representations.

The relationship between configuration spaces and operads can be explained in the following way. For technical reasons, we will restrict our attention to framed manifolds, i.e., manifolds equipped with a trivialization of the tangent bundle. Let us now consider a "fattened" version of configuration spaces, where points are replaced by embedded framed disks with pairwise disjoint interiors. The spaces of all configurations of such disks, denoted $\mathsf{D}_M(r)$, for a framed manifold M and an integer $r \geq 0$, are homotopy equivalent to the configuration spaces $\mathsf{Conf}_M(r)$ (see Sect. 5.3 for more precise definitions). The homotopy equivalence $\mathsf{D}_M(r) \to \mathsf{Conf}_M(r)$ is simply the map which sends a configuration of disks to the configuration of points formed by the centers of the disks.

These configuration spaces of disks have an extra structure that does not directly appear on configuration spaces of points. Consider the unit disk $\mathbb{D}^n$. The collection of spaces $\mathsf{D}_n := \mathsf{D}_{\mathbb{D}^n} = \{\mathsf{D}_{\mathbb{D}^n}(r)\}_{r \geq 0}$ is an operad using the following structure maps. Given two configurations $c = (c_1, \ldots, c_r) \in \mathsf{D}_n(r)$ and $d = (d_1, \ldots, d_s) \in \mathsf{D}_n(s)$, where the $c_i, d_j : \mathbb{D}^n \hookrightarrow \mathbb{D}^n$ are framed embeddings, we can define new configurations $c \circ_i d \in D_n(r + s - 1)$ (for $1 \le i \le r$) by using composition of embeddings (see Fig. 5.2):

$$c \circ_i d = (c_1, \ldots, c_{i-1}, c_i \circ d_1, \ldots, c_i \circ d_s, c_{i+1}, \ldots, c_r) \in D_n(r+s-1). \quad (1.1)$$

These composition maps satisfy several axioms: the sequential composition axiom, the parallel composition axiom, unitality, and equivariance with respect to the action of symmetric groups. Moreover, if M is a framed manifold, then the collection $D_M := \{D_M(r)\}_{r \ge 0}$ has the structure of a right module over the operad D_n. Given two configurations $c \in D_M(r)$ and $d \in D_n(s)$, we can define new configurations $c \circ_i d \in D_M(r + s - 1)$ ($1 \le i \le r$) by formulas analogous to the previous ones.

It turns out that the proofs of Kontsevich [Kon99] and Lambrechts and Volić [LV14] of the formality of the spaces $\mathrm{Conf}_{\mathbb{R}^n}(r) \simeq \mathrm{Conf}_{\mathbb{D}^n}(r) \simeq D_n(r)$ was actually about more than just the formality of the spaces on their own. In fact, their proofs show that the operad D_n itself is formal: the cohomology $H^*(D_n)$ is a model of the operad D_n in the sense of rational homotopy theory. The formality of the operad has profound consequences, for example the Deligne conjecture [KS00; MS02] or the deformation quantization of Poisson manifolds [Kon99; Kon03; Tam98].

In Chap. 5, after having introduced the theory of operads, we will explain how the results of the previous chapters interact with the operadic structures mentioned above. We will also show that our models (in graph complexes or à la Lambrechts–Stanley) have algebraic structures which mirror these operadic structures, and how our proofs are all compatible with these structures (see Sect. 5.4). We will finally conclude with an example of application of this to the computation of factorization homology (see Sect. 5.5), an invariant of manifolds defined from configuration spaces.

1.4.1 Conventions

We generally work with cohomologically-graded cochain complexes over $\mathbb{R}$, i.e., graded vector spaces $V = \{V^n\}_{n \in \mathbb{Z}}$ equipped with a differential of degree $+1$, $d : V^n \to V^{n+1}$. We also consider the (de)suspensions, for $k \in \mathbb{Z}$, to be defined by $(V[k])^n = V^{n+k}$.

Chapter 2
Configuration Spaces of Manifolds

In this chapter, we introduce configuration spaces of manifolds more precisely. We start by listing several of their applications and occurrences in mathematics in Sect. 2.1, including braid groups and surface braid groups, Goodwillie–Weiss calculus, Gelfand–Fuks cohomology, stable splitting of mapping spaces, iterated loop spaces, and stability.

After some quick background on homotopy theory, we explain the homotopy invariance conjecture in Sect. 2.2, which states that two homotopy equivalent manifolds have homotopy equivalent configuration spaces. We also explain why this conjecture must be refined, and in particular we explain the counterexample of Longoni and Salvatore [LS05] for non-simply connected closed manifolds. We also give a brief recollection of Sullivan's rational homotopy theory in Sect. 2.3. We explain how the conjecture can be adapted and strengthened in this setting: instead of merely asking for rational homotopy invariance, we will ask whether it is possible to actually compute the rational homotopy type of a configuration space in the form of a model.

Finally, we study the basic case of configuration spaces of Euclidean spaces in Sect. 2.4. Since all manifolds are built by gluing Euclidean spaces, it is natural to start there. As proved by Arnold [Arn69] (for $n = 2$) and Kontsevich [Kon99] and Lambrechts and Volić [LV14] (for all n), the configuration spaces of Euclidean spaces are formal, which allows one to make many explicit computations. The proof of that formality (which is actually postponed until Chap. 5 for $n \neq 2$) serves as the basis for the proofs in the later chapters.

2.1 Configuration Spaces

Let M be a topological manifold and $r \geq 0$ an integer. The following object will be the central object of study in these notes.

© The Author(s), under exclusive license to Springer Nature Switzerland AG 2022
N. Idrissi, *Real Homotopy of Configuration Spaces*, Lecture Notes
in Mathematics 2303, https://doi.org/10.1007/978-3-031-04428-1_2

Definition 2.1 The *rth configuration space* of M is the subspace of M^r given by:

$$\mathsf{Conf}_M(r) := \big\{(x_1, \ldots, x_r) \in M^r \mid \forall i \neq j,\ x_i \neq x_j\big\}. \tag{2.1}$$

It is topologized as an open submanifold of M^r.

A point of $\mathsf{Conf}_M(r)$ is given by an ordered collection of r pairwise distinct points in M. In order to emphasize the various aspects of this definition, these configuration spaces are sometimes called the *ordered* configuration spaces, or the configuration spaces *of points*, or a combination of both. These spaces have interesting topology, as we will illustrate below. Before moving on to more complex examples, let us first illustrate the simplest cases of configuration spaces.

Example 2.2 The configuration space $\mathsf{Conf}_M(0)$ is simply a singleton, given by the empty configuration. The space $\mathsf{Conf}_M(1)$ is just M itself.

Example 2.3 The space $\mathsf{Conf}_M(2)$ is given by $\{(x, y) \in M^2 \mid x \neq y\}$. In other words, it is the square of M with the diagonal $\{(x, x) \in M^2\}$ removed.

As r increases, the spaces $\mathsf{Conf}_M(r)$ become more and more complicated.

Example 2.4 Let $M = \mathbb{R}$ be the real line. Then the space $\mathsf{Conf}_{\mathbb{R}}(r)$ has $r!$ connected components. Each connected component is represented by a way of ordering the different points in $\mathbb{R}$.

There are numerous variants of configuration spaces. One example is given by unordered configuration spaces, which are defined as follows.

Definition 2.5 Let M be a manifold and $r \geq 0$ be an integer. There is an action of the symmetric group $\mathfrak{S}_r$ on $\mathsf{Conf}_M(r)$ by renumbering. A permutation $\sigma \in \mathfrak{S}_r$ simply acts by $(x_1, \ldots, x_r) \cdot \sigma = (x_{\sigma(1)}, \ldots, x_{\sigma(r)})$. The *unordered configuration space* is the orbit space:

$$\mathsf{UConf}_M(r) := \mathsf{Conf}_M(r)/\mathfrak{S}_r, \tag{2.2}$$

i.e., we consider subsets of M of cardinality exactly r.

Configuration spaces appear in many contexts. These spaces have been studied since at least the 1960's. To the author's knowledge, the name "configuration space" was coined by Fadell and Neuwirth [FN62a]. Let us now mention a few examples of occurrences of configuration spaces in the literature. This list is neither exhaustive nor chronological; its main purpose is to give the curious reader pointers to the literature.

Example 2.6 Let $r \geq 0$ be an integer. The *braid group on r strands* (also called the Artin braid group after Artin [Art47]) is the group with the following presentation:

$$B_r := \langle s_1, \ldots, s_{r-1} \mid s_i s_j = s_j s_i \text{ for } |i - j| \geq 2,\ s_i s_{i+1} s_i = s_{i+1} s_i s_{i+1} \rangle. \tag{2.3}$$

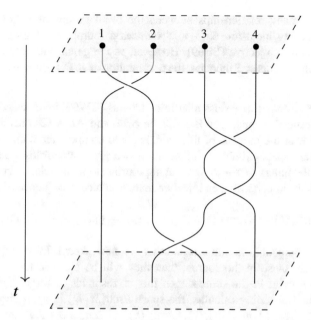

Fig. 2.1 The braid $s_1 s_3 s_3 s_2^{-1}$ seen as a path in $\mathsf{Conf}_{\mathbb{R}^2}(4)$

There is a well-defined morphism from B_r to the symmetric group Σ_r that sends s_i to the transposition that exchanges i and $i + 1$. Its kernel is called the *pure braid group on r strands* and is denoted PB_r.

Fox and Neuwirth [FN62b] proved that B_r is isomorphic to the fundamental group of the unordered configuration space of r points in the plane, i.e., $PB_r \cong \pi_1(\mathsf{UConf}_{\mathbb{R}^2}(r))$. A braid simply corresponds to a path in the configuration space, and an isotopy corresponds to a homotopy. We refer to Fig. 2.1 for an illustration. The pure braid group B_r is the fundamental group of the ordered configuration space, i.e., $PB_r \cong \pi_1(\mathsf{Conf}_{\mathbb{R}^2}(r))$. An element of PB_r corresponds to a path in the configuration space such that each strand starts and ends at the same position.

The configuration spaces of the plane are aspherical, i.e., their universal covers are trivial. It follows that $\mathsf{Conf}_{\mathbb{R}^2}(r)$ (resp. $\mathsf{UConf}_{\mathbb{R}^2}(r)$) is the classifying space of the group B_r (resp. PB_r). We can thus compute, for example, the cohomology – in the algebraic sense– of the braid groups by computing the cohomology –in the topological sense– of the configuration spaces of the plane.

Example 2.7 The surface braid groups are a generalization of the previous example. Given a surface S (i.e., a compact manifold of dimension 2, possibly with boundary) and an integer $r \geq 0$, the configuration space $\mathsf{Conf}_S(r)/\mathfrak{S}_r$ is aspherical, just like $\mathsf{Conf}_{\mathbb{R}^2}(r)$. Its fundamental group $B_r(S)$ is called the braid group of the surface S on r strands. The corresponding pure braid group $PB_r(S)$ is $\pi_1(\mathsf{Conf}_S(r))$. We refer to [GJ15] for a survey on the subject. Fadell and Neuwirth [FN62a, Theorem 9] have proved that surfaces are the only case of interest: higher-dimensional manifolds do not admit braid theories.

There are many relationships between the braid groups of various surfaces. For example, any inclusion $\mathbb{R}^2 \rightarrow S$ induces a morphism of groups $PB_r \rightarrow PB_r(S)$, which is injective [Bir69]. However, as the genus and/or the number of boundary components of S increase, the presentation of $B_r(S)$ becomes increasingly complicated.

Example 2.8 Goodwillie–Weiss manifold calculus [GW99] involves configuration spaces in a crucial manner (see Boavida de Brito and Weiss [BW18] and Turchin [Tur13]). One of the purpose of this calculus is to compute the homotopy type of the embedding spaces Emb(M, N) between two given manifolds, defined as the set of all embeddings $M \hookrightarrow N$ with an appropriate topology. Since an embedding $f : M \hookrightarrow N$ is, in particular, an injective map, it induces a sequence of maps,

$$f_k : \mathsf{Conf}_M(k) \rightarrow \mathsf{Conf}_N(k), \quad (x_1, \ldots, x_k) \mapsto (f(x_1), \ldots, f(x_k)), \qquad (2.4)$$

that are all compatible in some sense that we now sketch. Most importantly, if two points are close in the source, then they will be close in the target, and if one forgets a point in the source, then this corresponds to forgetting a point in the target. In embedding calculus, the space Emb(M, N) is approximated as the subspace of $\prod_{k \in \mathbb{N}} \mathsf{Map}(\mathsf{Conf}_M(k), \mathsf{Conf}_N(k))$ of sequences that are compatible –in the previous sense– up to homotopy. Under good conditions (in particular $\dim N - \dim M \geq 3$), this process allows one to recover the homotopy type of the space Emb(M, N). (To be completely precise, one can recover the homotopy type of the space of *framed* embeddings $\mathsf{Emb}^{\mathrm{fr}}(M, N)$ between two framed manifolds; to recover Emb(M, N), one needs to use framed configuration spaces, see Sect. 5.3.2.)

Example 2.9 Let M be a smooth manifold and let $\Gamma_c(M, TM)$ be the Lie algebra (see Definition 2.77) of vector fields with compact support on M. The Gelfand–Fuks cohomology of M, given $H^*_{\mathrm{cont}}(\Gamma_c(M, TM))$ (where the cohomology is a continuous version of the Chevalley–Eilenberg cochains, see Definition 5.104), is an invariant that appears in the study of characteristic classes of foliations see Morita [Mor01].

Cohen and Taylor [CT78] built a spectral sequence which converges to Gelfand–Fuks cohomology of M and whose E^2 page can be expressed in terms of (decorated) configuration spaces of M. The relationship between Gelfand–Fuks cohomology and configuration spaces can be seen as stemming from the following description of the cochain complex of $\Gamma_c(M, TM)$ (see Haefliger [Hae76, Section 5] for details). Consider the direct limit $\mathsf{Conf}_M := \varprojlim_{r \geq 0} \mathsf{Conf}_M(r)$, with a topology which is roughly speaking given by allowing points to vanish at infinity. Then the continuous Chevalley–Eilenberg cochain complex $C^*_{\mathrm{cont}}(\Gamma_c(M, TM))$ is isomorphic to the space of sections of the differentiable sheaf on Conf_M given by continuous linear forms (with compact support) on the cosheaf of vector fields.

Example 2.10 (Snaith [Sna74]) The space of maps with compact support

$$\mathrm{Map}_c(\mathbb{R}^d, Y) := \{f : \mathbb{R}^d \to Y \mid \exists K \subset \mathbb{R}^d \text{ compact s.t. } f|_{\mathbb{R}^d \setminus K} \text{ is constant}\}$$
$$(2.5)$$

splits stably as a wedge sum of terms given by decorated configuration spaces of $\mathbb{R}^d$. More precisely, if X is a compact manifold with boundary, then there is a stable homotopy equivalence (i.e., a map which induces an isomorphism on homotopy once enough the suspension functor has been applied enough times) of the form:

$$\mathrm{Map}_c(\mathbb{R}^d, \Sigma^d(X/\partial X)) \underset{\mathrm{st}}{\simeq} \bigvee_{n \geq 0} \mathrm{Conf}_{\mathbb{R}^d}(n; X).$$
$$(2.6)$$

Example 2.11 In classical mechanics, the configuration spaces we have been writing about are a particular case of the more general notion of the configuration space of a physical system. We are essentially considering the case of discrete particles that cannot occupy the same position at the same time. We refer to Axelrod and Singer [AS92] for explanations of the relationship with the physics literature.

Example 2.12 Configuration spaces are closely related to iterated loop spaces. We will give more detail in Chap. 5 (and particularly in the discussion surrounding Definition 5.41) as this connection has to do with operads. Let us just give the following intuitive picture, that we reproduce from Segal [Seg73, Section 1]. Let $\mathbb{S}^n = \mathbb{R}^n \cup \{\infty\}$ be the sphere. There is a map from $\mathrm{Conf}_{\mathbb{R}^n}(r)$ to the space of maps $\mathbb{S}^n \to \mathbb{S}^n$ which preserve the base point (i.e., $\Omega^n \mathbb{S}^n$ in the language of Definition 5.41). This map has a description inspired from physics. We can view a configuration $(x_1, \ldots, x_r) \in \mathrm{Conf}_{\mathbb{R}^n}(r)$ as a collection of electrically charged particles, each with a unit charge. These particles generate an electric field, which is a vector field on $\mathbb{R}^n \setminus \{x_i\}$, or equivalent a smooth map $E_x : \mathbb{R}^n \setminus \{x_i\} \to \mathbb{R}^n$. This map E_x can be extended as a map $E_x : \mathbb{S}^n \to \mathbb{S}^n$ by setting $E_x(x_i) = \infty$ and $E_x(\infty) = 0$.

Example 2.13 (Motion Planning) Assume that we are given a number of robots moving around in a space M and all controlled by a central computer. We wish to move all the robots at the same time between a starting and a finishing position (Fig. 2.2). The robots cannot be at the same location at the same time, so we need to find a path inside the configuration space $\mathrm{Conf}_M(r)$, where r is the number of robots. More precisely, let us denote the space of all paths in $\mathrm{Conf}_M(r)$ by $P\mathrm{Conf}_M(r) = \mathrm{Map}([0, 1], \mathrm{Conf}_M(r))$ There is a canonical projection defined by:

$$p : P\mathrm{Conf}_M(r) \to \mathrm{Conf}_M(r) \times \mathrm{Conf}_M(r)$$
$$\gamma \mapsto (\gamma(0), \gamma(1)).$$
$$(2.7)$$

We are thus looking for a section of p, i.e., a map $\sigma : \mathrm{Conf}_M(r) \times \mathrm{Conf}_M(r) \to P\mathrm{Conf}_M(r)$ that takes a couple of configurations and returns a path between them. Unless $\mathsf{Conf}_M(r)$ is contractible, which rarely happens, such a section can never be continuous. There exists an invariant, called the topological complexity $\mathrm{TC}(\mathsf{Conf}_M(r))$, which is defined as the minimum number of continuity domains of such a section (cf. Farber [Far03]). This invariant only depends on the homotopy type of $\mathsf{Conf}_M(r)$. It is related to the more classical invariant called Lusternik–Schnirelmann category $\mathrm{cat}(\mathsf{Conf}_M(r))$ [LS34], i.e., the minimal number (+1) of contractible open subsets required to cover $\mathsf{Conf}_M(r)$. One has

$$\mathrm{cat}(\mathsf{Conf}_M(r)) \leq \mathrm{TC}(\mathsf{Conf}_M(r)) \leq 2 \cdot \mathrm{cat}(\mathsf{Conf}_M(r)) - 1 \qquad (2.8)$$

thanks to [Far03, Theorem 5].

Remark 2.14 Configuration spaces of topological spaces that are not manifolds are also objects of interest. For example, there is a growing body of literature about configuration spaces of graphs, see for example [ADK19]. They can be of particular interest in e.g., robotics, see Ghrist [Ghr01] and Farber [Far18].

Example 2.15 (Stability) It has long been known that the homology of the unordered configuration spaces (Definition 2.5) stabilizes. This was first proved by Arnold [Arn69] for the plane: there are homotopy classes of maps $\mathsf{UConf}_{\mathbb{R}^2}(r) \to \mathsf{UConf}_{\mathbb{R}^2}(r+1)$ (intuitively given by "adding a point at infinity") that induce an isomorphism in homology for r large enough compared to the homological degree. This was later generalized for open manifolds by McDuff [McD75] and Segal [Seg79]. However, it is relatively easy to prove that the homology of *ordered* configuration spaces is not stable (Corollary 2.90 implies that $\dim H_1(\mathsf{Conf}_{\mathbb{R}^2}(r)) = \binom{r}{2}$ is not eventually constant, for example).

Church [Chu12] has proved that the homology of the ordered configuration spaces of a closed manifold *does* stabilize in a different sense. If one decomposes the homology of $\mathsf{Conf}_M(r)$ as a direct sum of irreducible representations of the symmetric group $\mathfrak{S}_r$, then there is a canonical (standard) way of indexing the representations which eventually stabilizes. This immediately implies the stability of the homology of the unordered configuration space, as this homology corresponds to the summand given by the trivial representation. This phenomenon, called representation stability (see Djament and Vespa [DV10] and Church and Farb [CF13]), is the subject an active area of research, of which we can hardly give a full account.

Fig. 2.2 Motion planning: how to find non-intersecting paths that depend continuously on the starting and ending positions?

Let us just note the following, without going into detail. Representation stability has been observed by Church et al. [CEF15] to be the consequence of another result. The collection $\mathsf{Conf}_M = \{\mathsf{Conf}_M(r)\}_{r \geq 0}$ has the structure of an $\mathsf{FI}^{\mathrm{op}}$-module, where FI is the category of finite sets and injections; concretely, this means that one can renumber and/or forget points in a configuration. The cohomology of Conf_M in a given degree thus forms an FI-module by functoriality. It turns out that this FI-module is finitely generated (i.e., it contains a finite subspace S such that the smallest FI-module containing S is the whole FI-module). Finitely generated FI-modules automatically satisfy representation stability thanks to the result of [CEF15]. This has an interpretation in terms of strongly polynomial functors (see Djament and Vespa [DV17]). A strongly polynomial functor $F : \mathcal{M} \to \mathcal{A}$ from a monoidal category $\mathcal{M}$ to an abelian category $\mathcal{A}$ is a functor which satisfies a condition analogous to the characterization of polynomial maps $f : \mathbb{R} \to \mathbb{R}$ as those such that the iterated derivatives $\{f^{(n)}\}_{n \geq 0}$ become identically zero at a finite stage (the degree of the polynomial $+1$). The result of [CEF15] can be reinterpreted as the fact that a finitely generated FI-module is a strongly polynomial functor. Let us finally note that the FI-module structure is part of the natural action of the little disks operad on configuration spaces, see Chap. 5 (forgetting a point is just inserting an empty configuration).

There are many generalizations of configuration spaces. Let us list a couple.

Example 2.16 Let $k \geq 2$. Given a manifold M and an integer $r \geq 0$, one can define the non-k-equal configuration space of r points on M by:

$$\mathsf{Conf}_M^{<k}(M) := \{(x_1, \ldots, x_r) \in M^r \mid \nexists i_1 < \cdots < i_k \text{ s.t. } x_{i_1} = \cdots = x_{i_k}\}. \quad (2.9)$$

In other words, the space $\mathsf{Conf}_M^{<k}(r)$ is the space of collections of r points in M such that at most $k - 1$ points can collide at once. Of course, the space of non-2-equal configurations, $\mathsf{Conf}_M^{<2}(r)$ is simply the usual configuration space $\mathsf{Conf}_M(r)$. By contrast, for $r < k$, one gets $\mathsf{Conf}_M^{<k}(r) = M^r$. As soon as $k \geq 3$, then the fundamental group of $\mathsf{Conf}_M^{<k}(r)$ becomes abelian, unlike the braid groups (see Kallel and Saihi [KS16]). These spaces have also applications in Goodwillie–Weiss embedding calculus: instead of computing spaces of embeddings, they can be used to compute spaces of non-k-equal immersions (i.e., immersions $f : M \to N$ such that the preimages $f^{-1}(y)$ have cardinality at most $k - 1$ for all $y \in N$), see Dobrinskaya and Turchin [DT15] and Grossnickle [Gro19].

Example 2.17 Let Γ be a simple graph (i.e., a graph without loops or double edges) on with set of vertices $\{1, \ldots, r\}$, where $r \geq 0$ is some integer. Then one can define the *generalized configuration space* $\mathsf{Conf}_M(\Gamma)$ by:

$$\mathsf{Conf}_M(\Gamma) := \{(x_1, \ldots, x_r) \in M^r \mid (i, j) \in \Gamma \implies x_i \neq x_j\}. \quad (2.10)$$

The classical configuration space $\mathsf{Conf}_M(r)$ simply corresponds to the generalized configuration space associated to the complete graph on r vertices. The cohomology of such configuration spaces has been computed by Petersen [Pet20] and Bökstedt and Minuz [BM20].

2.2 Homotopy Invariance

In all of the applications mentioned above, knowing the homotopy type of $\text{Conf}_M(r)$ is essential. To set the stage for rational homotopy theory and fix notation, let us now quickly recall the basics of homotopy theory. All the maps we consider below are assumed to be continuous.

Definition 2.18 Let A and X be two spaces. Two maps $f, g : A \to X$ are *homotopic* if there exists a map $H : A \times [0, 1] \to X$ (called a homotopy) such that $H(-, 0) = f$ and $H(-, 1) = g$. If this is the case, we denote it by $f \simeq g$.

Definition 2.19 A map $f : A \to X$ is a *homotopy equivalence* if there exists a map $g : X \to A$ such that $f \circ g \simeq \text{id}_X$ and $g \circ f \simeq \text{id}_A$. If such a map exists, we say that A and X have the same *homotopy type*, or that they are *homotopy equivalent*.

Example 2.20 The real line $\mathbb{R}$ is homotopy equivalent to a singleton. More generally, any Euclidean space $\mathbb{R}^n$ is homotopy equivalent to a singleton.

Let us now consider configuration spaces. A very natural question is the following: if two spaces are homotopy equivalent, are their configuration spaces then homotopy equivalent? In other words, do we have

$$M \simeq N \overset{?}{\implies} \text{Conf}_M(r) \simeq \text{Conf}_N(r) \tag{2.11}$$

This is far from obvious. Indeed, nothing indicates that a homotopy equivalence is injective, so it may not even define a map between configuration spaces. Even if some homotopy equivalences $f : M \leftrightarrows N : g$ are injective (and thus induce maps $\text{Conf}_M(r) \leftrightarrows \text{Conf}_N(r)$), the homotopies $g \circ f \simeq \text{id}_M$ and $f \circ g \simeq \text{id}_N$ may not necessarily be injective at all times, so they may not induce a homotopy at the level of configuration spaces.

In fact, the question asked above is too naive. There is an obvious counterexample, given by Example 2.20. Indeed, the space $\text{Conf}_{\{0\}}(2)$ is empty, because it is not possible to find two distinct points in a singleton. However, $\text{Conf}_{\mathbb{R}}(2)$, i.e., the plane with the diagonal removed, is clearly nonempty. Therefore these two configuration spaces are not homotopy equivalent. More generally, we have the following result, from which it follows that the homotopy types of the spaces $\text{Conf}_{\mathbb{R}^n}(2)$ are pairwise distinct.

Lemma 2.21 *The configuration space* $\text{Conf}_{\mathbb{R}^n}(2)$ *is homotopy equivalent to the* $(n-1)$-*sphere* $\mathbb{S}^{n-1}$.

Proof View the sphere $\mathbb{S}^{n-1}$ as the following space:

$$\mathbb{S}^{n-1} := \{x \in \mathbb{R}^n \mid \|x\| = 1\}. \tag{2.12}$$

Then we can define an "angle map", that records the angle between two points in a configuration:

$$\theta : \mathsf{Conf}_{\mathbb{R}^n}(2) \to \mathbb{S}^{n-1}, \quad (x, y) \mapsto \frac{x - y}{\|x - y\|}. \tag{2.13}$$

Conversely, we have a map:

$$i : \mathbb{S}^{n-1} \to \mathsf{Conf}_{\mathbb{R}^n}(2) \quad v \mapsto (v, -v). \tag{2.14}$$

Intuitively, the map i describes the rotation of two points around a common center of mass located at the origin, with a unit radius. We can immediately see that $\theta \circ i = \mathrm{id}_{\mathbb{S}^{n-1}}$. Moreover, the following describes a homotopy between $i \circ \theta$ and the identity of $\mathsf{Conf}_{\mathbb{R}^n}(2)$:

$$H : \mathsf{Conf}_{\mathbb{R}^n}(2) \times [0, 1] \to \mathsf{Conf}_{\mathbb{R}^n}(2)$$

$$((x, y), t) \mapsto \left(\frac{\|x - y\|}{2}\right)^{-t} \cdot \left(u(x, y, t), \, v(x, y, t)\right) \tag{2.15}$$

where $u(x, y, t) = x - \frac{t}{2}(x + y)$ and $v(x, y, t) = y - \frac{t}{2}(x + y)$. $\qquad\square$

Remark 2.22 In fact, the space $\mathsf{Conf}_{\mathbb{R}^n}(2)$ is homeomorphic to the product $\mathbb{S}^{n-1} \times \mathbb{R}^n \times (0, +\infty)$. The homeomorphism $\mathsf{Conf}_{\mathbb{R}^n}(2) \to \mathbb{S}^{n-1} \times \mathbb{R}^n \times (0, +\infty)$ is the product of the map θ with $(x, y) \mapsto (x + y)/2$ and $(x, y) \mapsto \|x - y\|$. The spaces $\mathsf{Conf}_{\mathbb{R}^n}(r)$ for higher r have a more complicated homotopy type. They form a special case of what is known as an hyperplane arrangement complement. We refer to Orlik and Terao [OT92] for an introduction of the subject.

The counterexample of Lemma 2.21 for the homotopy invariance conjecture mainly stems from the fact that we are considering manifolds of differing dimensions. However, even if one restricts the question to manifolds of equal dimension, counterexamples still exist, as shown by the following.

Example 2.23 ([Knu18a, Section 1.1]) Let $\Sigma_{1,1} = (\mathbb{S}^1 \times \mathbb{S}^1) \setminus \{*\}$ be the torus with a point removed, and let $\Sigma_{0,3} = \mathbb{R}^2 \setminus \{*, *'\}$ be the plane with two points removed. Then $\Sigma_{1,1} \simeq \Sigma_{0,3} \simeq \mathbb{S}^1 \vee \mathbb{S}^1$ are homotopy equivalent, since they admit the figure-eight graph (i.e., a wedge sum of two circles) as a deformation retract. However, the space $\mathsf{Conf}_{\Sigma_{1,1}}(2)$ is not homotopy equivalent to $\mathsf{Conf}_{\Sigma_{0,3}}(2)$. This counterexample can be proved with surface braid groups (see Example 2.7). Recall that $PB_2(\Sigma_{g,n})$ is the fundamental group of $\mathsf{Conf}_{\Sigma_{g,n}}(2)$, where $\Sigma_{g,n}$ is the closed oriented surface of genus g with n holes. An element of $PB_2(\Sigma_{0,3})$ is a (pure) braid with four strands such that the last two braids remain in a fixed position. An element of $PB_2(\Sigma_{1,1})$, however, is a (pure) braid with three strands on a torus, where the third point remains in a fixed position, and the two other strands can move along the meridian and the longitude of the torus in addition to crossing over and under other points. These groups have a presentation given in [Lam00] for $g = 0$ and [Bel04, Theorem 5.2] for $g > 0$. In these presentations, we see that both $PB_2(\Sigma_{0,3})$ and $PB_2(\Sigma_{1,1})$ are generated by five elements, denoted by $A_{1,3}, A_{2,3}, A_{1,4}, A_{2,4}, A_{3,4}$. However, in $PB_2(\Sigma_{1,1})$, one of the generators is redundant (see [Bel04, Remark 5.5]): the pure braid where the two points switch places twice is a commutator of the two braids where the first point goes along the meridian (resp. the longitude). It follows that the abelianization of $PB_2(\Sigma_{0,3})$ has rank 5, whereas that of $PB_2(\Sigma_{1,1})$ has rank 4.

This new counterexample essentially stems from the fact that the manifolds considered are open (in this case, they are not compact, but if we decide to compactify them then we must add a nonempty boundary). It is thus natural to restrict the question even further to closed manifolds, i.e., compact manifolds without boundary.

Up to dimension 2, it is clear that homotopy equivalent closed manifolds have homotopy equivalent configuration spaces. Indeed, in dimension 0 and 1, there are only two closed manifolds up to homeomorphism (namely $\{0\}$ and $\mathbb{S}^1$). In dimension 2, two homotopy equivalent surfaces are always homeomorphic, thanks to the classification of surfaces. It is moreover clear that homeomorphic spaces have homeomorphic (and thus homotopy equivalent) configuration spaces.

Several results led to believe that the homotopy invariance of configuration spaces of closed manifolds was true in higher dimension:

- Results of Bödigheimer et al. [BCT89] and Bendersky and Gitler [BG91] showed that the homology of $\mathsf{Conf}_M(r)$ only depends on the homotopy type of M (up to certain hypotheses on the dimension of M or the characteristic of the base field).
- Levitt [Lev95] showed that the homotopy type of the loop space $\Omega\mathsf{Conf}_M(r)$ of the configuration space of M only depends on the homotopy type of M. In particular, this shows that the homotopy groups of $\mathsf{Conf}_M(r)$ only depend on the homotopy type of M.
- Aouina and Klein [AK04] proved that the stable homotopy type of $\mathsf{Conf}_M(r)$ is also a homotopy invariant, i.e., if $M \simeq N$ then $\Sigma^k\mathsf{Conf}_M(r) \simeq \Sigma^k\mathsf{Conf}_N(r)$ for $k \gg 0$, where $\Sigma X = (-) \wedge \mathbb{S}^1$ is the suspension.

However, a counterexample was found some years ago by Longoni and Salvatore [LS05]. This counterexample is given by lens spaces, that we now define.

Definition 2.24 Let p, q be coprime integers. The *lens space* $L_{p,q}$ is the orbit space of $\mathbb{S}^3 = \{(z, z') \in \mathbb{C}^2 \mid |z|^2 + |z'|^2 = 1\}$ under the action of the cyclic group $\mathbb{Z}/p\mathbb{Z}$ given by:

$$(z, z') \mapsto (e^{2i\pi/p}z, e^{2iq\pi/p}z') \tag{2.16}$$

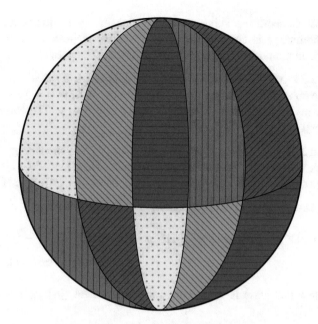

Fig. 2.3 Identifications of triangles in $\partial \mathbb{D}^3 = \mathbb{S}^2$ to obtain $L_{5,2}$

Since p and q are coprime, this action is free and the quotient $L_{p,q}$ is a three-dimensional closed manifold.

Example 2.25 The lens space $L_{2,1}$ is the real projective space $\mathbb{RP}^3$.

Remark 2.26 Alternatively, one can see $L_{p,q}$ as a quotient of the ball $\mathbb{D}^3$, see Fig. 2.3. Indeed, let us start by cutting the boundary $\partial \mathbb{D}^3 = \mathbb{S}^2$ into $2p$ triangles: the first p triangles start from the north pole, the last p from the south pole, and they all join together in pairs at the equator (divided into p segments). To get $L_{p,q}$, one then identifies the ith northern triangle with the $(i + q)$th southern triangle.

Theorem 2.27 (Longoni and Salvatore [LS05]) *The two closed manifolds $L_{7,1}$ and $L_{7,2}$ have the same homotopy type but their configuration spaces are not homotopy equivalent.*

The proof of Longoni and Salvatore of the fact that $\mathrm{Conf}_{L_{7,1}}(r) \not\simeq \mathrm{Conf}_{L_{7,2}}(r)$ for $r \geq 2$ involves Massey products.

Notation 2.28 We will denote the homology of a space X with coefficients in a ring $\Bbbk$ by $H_*(X; \Bbbk) := \bigoplus_{n=0}^{\infty} H_n(X; \Bbbk)$. Cohomology will similarly be denoted by $H^*(X; \Bbbk) := \bigoplus_{n=0}^{\infty} H^n(X; \Bbbk)$. Usually, the ring will be implicit from the context (often $\mathbb{Q}$ or $\mathbb{R}$) and will be omitted from the notation.

Remark 2.29 Generally speaking, we will consider all degrees to be cohomological, so that cohomology is concentrated in nonnegative degrees while homology is concentrated in nonpositive degrees (with $(H_*(X; \Bbbk))^{-n} := H_n(X; \Bbbk)$).

Definition 2.30 (Massey [Mas58]) Let X be a topological space and let $[\alpha]$, $[\beta]$, $[\gamma]$ be three cohomology classes on X satisfying $[\alpha \smile \beta] = 0 = [\beta \smile \gamma]$. There exists cochains λ, μ such that $d\lambda = \alpha \smile \beta$ and $d\mu = \beta \smile \gamma$. The *Massey product* $\langle [\alpha], [\beta], [\gamma] \rangle$ is $[\lambda\gamma - (-1)^{|\alpha|}\alpha\mu]$.

Proposition 2.31 *Let X be a space and $[\alpha]$, $[\beta]$, $[\gamma]$ be cohomology classes and λ, μ be cochains as in the previous proposition. Then $[\lambda\gamma - (-1)^{|\alpha|}\alpha\mu]$ is a cohomology class which is well-defined up to the ideal $([\alpha], [\gamma]) \subset H^*(X)$.*

Proof It is straightforward to check that $\lambda\gamma - (-1)^{|\alpha|}\alpha\mu$ is a cocycle:

$$d(\mu\gamma - (-1)^{|\alpha|}\alpha\gamma) = d\mu \cdot \gamma + (-1)^{|\mu|}\mu \cdot d\gamma - (-1)^{|\alpha|}d\alpha \cdot \gamma - (-1)^{2|\alpha|}\alpha \cdot d\gamma$$

$$= \alpha\beta\gamma - \alpha\beta\gamma = 0. \qquad (2.17)$$

Checking that this class is well-defined up to the ideal $([\alpha], [\gamma])$ is a classical exercise. $\qquad\qquad\square$

Proof (Idea of the Proof of Theorem 2.27) Let us now quickly give an idea of the proof of the theorem of Longoni and Salvatore [LS05], which we will largely paraphrase.

It is clear that the universal cover of $L_{p,q}$ is always $\mathbb{S}^3$. The universal covering space $\widetilde{\mathrm{Conf}}_{L_{p,q}}(2)$ is thus given by:

$$\widetilde{\mathrm{Conf}}_{L_{p,q}}(2) = \{(x, y) \in \mathbb{S}^3 \times \mathbb{S}^3 \mid \forall k \in \mathbb{Z}/p\mathbb{Z},\ x \neq k \cdot y\}, \qquad (2.18)$$

where $\mathbb{Z}/p\mathbb{Z}$ acts on $\mathbb{S}^3$ through Eq. (2.16). Then we get the following description of $\widetilde{\mathrm{Conf}}_{L_{7,q}}(2)$ for $q \in \{1, 2\}$:

- Let us temporarily view $\mathbb{S}^3$ as the space of unitary quaternions, i.e., $\mathbb{S}^3 = U_1(\mathbb{H})$. Then the action of $\mathbb{Z}/7\mathbb{Z}$ is translation by $\mathbb{Z}/7\mathbb{Z} \subset \mathbb{C} \subset \mathbb{H}$. There is thus a homeomorphism $\widetilde{\mathrm{Conf}}_{L_{7,1}}(2) \cong (\mathbb{S}^3 \setminus \mathbb{Z}/7\mathbb{Z}) \times \mathbb{S}^3$ given by $(x, y) \mapsto (y^{-1}x, y)$. This space has the homotopy type as the wedge sum $(\mathbb{S}^2)^{\vee 6} \vee \mathbb{S}^3$. The Massey products of this space thus all vanish thanks to the fact this space is formal (see Definition 2.92 and the discussion afterwards).
- On the other hand, Longoni and Salvatore have found an explicit nonzero Massey product in $\widetilde{\mathrm{Conf}}_{L_{7,2}}(2)$. They define the isotopy (for $k \in \mathbb{Z}/7\mathbb{Z}$):

$$H_k : \mathbb{S}^3 \times [0, 1] \to \mathbb{S}^3 \times \mathbb{S}^3$$
$$((z_1, z_2), t) \mapsto \left((z_1, z_2), (e^{2i(k-1+t)\pi/7}z_1,\ e^{4i(k-1+t)\pi/7}z_2)\right). \qquad (2.19)$$

Then the image of H_k is a 4-submanifold $A_k \subset \mathbb{S}^3 \times \mathbb{S}^3$ which does not intersect the diagonals and thus represents, by Poincaré–Lefschetz duality, a cohomology class

$$a_k \in H^2(\mathrm{Conf}^{\sim}_{L_{7,2}}(2)). \tag{2.20}$$

These classes span the second cohomology group, and they are subject to the relation $\sum_k a_k = 0$. Longoni and Salvatore then determine how the submanifolds A_1, A_2, A_4, and A_6 intersect geometrically. They then use it to compute geometrically the Massey product $\langle [a_4], [a_1], [a_2 + a_5] \rangle$ and prove that it is nonzero.

This thus shows that $\mathrm{Conf}_{L_{7,1}}(2)$ and $\mathrm{Conf}_{L_{7,2}}(2)$ cannot have the same homotopy type, even though $L_{7,1} \simeq L_{7,2}$. A similar proof works to show that $\mathrm{Conf}_{L_{7,1}}(r) \not\simeq \mathrm{Conf}_{L_{7,2}}(r)$ for $r \geq 2$. $\square$

Despite this counterexample, there is still hope regarding the homotopy invariance conjecture. Indeed, the lens spaces are not simply connected: there exists loops that cannot be contracted to a point. In fact, their fundamental group is clearly given by $\pi_1(L_{p,q}) = \mathbb{Z}/p\mathbb{Z}$ (see Notation 2.35). Moreover, the proof of Longoni and Salvatore deals with Massey products in the universal cover, which needs the fundamental group to be nontrivial. The following conjecture thus remains open:

Conjecture 2.32 Let M and N be two simply connected closed manifolds. If M and N have the same homotopy type, then so do $\mathrm{Conf}_M(r)$ and $\mathrm{Conf}_N(r)$ for all $r \geq 0$.

Remark 2.33 Two simply connected closed manifolds have the same homotopy type if and only if they have the same *simple* homotopy type (i.e., they are related by a sequence of collapses and expansions of CW cells, see [Coh73b]). A more general conjecture thus states that if two closed manifolds have the same simple homotopy type, then so do their configuration spaces.

2.3 Rational Homotopy Invariance

In this volume, we will focus on the *rational* part of homotopy types (then, later on, on the *real* part). In summary, this means that we study homotopy types up to torsion. While this point of view loses information about spaces, it has the important advantage of being *computable*. The rational homotopy type of a space is completely described by some purely algebraic data called a (rational) model of the space. Let us now introduce this theory. In order to fix notation, we start again with some basic reminders on homotopy theory.

Remark 2.34 We cannot claim any kind of exhaustiveness about the field of rational homotopy theory in these notes. We refer to Félix et al. [FHT01; FOT08; FHT15], Hess [Hes07] and Griffiths and Morgan [GM13] for wider texts on the subject.

In order to motivate the definition of rational equivalences, let us first introduce the notion of weak homotopy equivalence.

Notation 2.35 For a topological space X, we let $\pi_0(X)$ (called the *0th homotopy group*) be the set of all path components of X. For $n \geq 1$ and $x_0 \in X$ a base point, $\pi_n(X, x_0)$ (called the *nth homotopy group*) is the group of all based homotopy classes of maps $(\mathbb{S}^n, *) \to (X, x_0)$.

Definition 2.36 A *weak homotopy equivalence* is a map $f : X \to Y$ such that $\pi_0(f)$ is a bijection and, for all $x_0 \in X$ and $n \geq 1$, $\pi_n(f, x_0)$ is an isomorphism.

We will denote weak equivalences with the symbol $\sim$ above the arrow, as in $f : X \xrightarrow{\sim} Y$.

Remark 2.37 A homotopy equivalence is obviously a weak homotopy equivalence.

Remark 2.38 If X and Y are related by a weak homotopy equivalence, then we say that they have the same weak homotopy type. Note that this is not symmetric: a weak homotopy equivalence $f : X \to Y$ does not necessarily admit a weak inverse. To obtain an equivalence relation on topological spaces, it is thus necessary to consider zigzags of weak homotopy equivalences. More precisely, we will say that X and Y have the same weak homotopy type if there exists a diagram of the form:

$$X \xleftarrow{\sim} X_1 \xrightarrow{\sim} X_2 \xleftarrow{\sim} \ldots \xrightarrow{\sim} Y. \tag{2.21}$$

Theorem 2.39 (Whitehead [Whi49]) *If X and Y are two CW-complexes, then they have the same homotopy type if and only if they have the same weak homotopy type.*

Any manifold is homotopy equivalent to a CW-complex and thus the previous theorem applies to manifolds. Since we will generally focus on manifolds, we will generally omit the adjective "weak" when referring to homotopy equivalences.

Definition 2.40 A space X is said to be *simply connected* if it is path-connected and $\pi_1(X, x_0) = 0$ for any base point.

From now on, we will focus on simply connected spaces. Rational homotopy theory can be worked out in a more general setting. The "classical" theory applies to nilpotent spaces of finite type (see Sullivan [Sul77] for the foundations and Félix et al. [FHT01] for a reference). Recently, the theory was extended to path-connected spaces by Félix et al. [FHT15]. The case of simply connected spaces is simpler, and the theorems we need are easier to state in this setting. For example, we do not need to concern ourselves with base points, and we will generally drop them from the notation of homotopy groups.

Definition 2.41 A topological space X is said to be of *finite $\Bbbk$-type* (or simply *finite type* if the field is understood) if the $\Bbbk$-modules $H_n(X; \Bbbk)$ are finitely generated for all $n \geq 0$.

Example 2.42 A closed manifold is of finite type over $\mathbb{Z}$.

The higher homotopy groups $\pi_n(X)$ of a space X are abelian, so they can all be seen as $\mathbb{Z}$-modules. The following definition, which is inspired by Definition 2.36, is at the basis of rational homotopy theory.

Definition 2.43 Let X and Y be two simply connected spaces of finite $\mathbb{Q}$-type. A *rational (homotopy) equivalence* from X to Y is a map $f : X \to Y$ such that $\pi_n(f) \otimes_{\mathbb{Z}} \mathbb{Q}$ is an isomorphism for all $n \geq 2$.

Example 2.44 A map $\mathbb{S}^n \to \mathbb{S}^n$ of degree k is not a weak equivalence unless $k = \pm 1$, but it is a rational equivalence as soon as $k \neq 0$.

Definition 2.45 We say that X and Y have the same *rational homotopy type* if they are connected by a zigzag of rational equivalences (see Remark 2.38) and we denote this by $X \simeq_{\mathbb{Q}} Y$.

Theorem 2.46 (Serre [Ser53]) *Let X and Y be two simply connected spaces of finite type and let $f : X \to Y$ be a map. Then f is a rational equivalence if and only if $H_*(f; \mathbb{Q})$ is an isomorphism, if and only if $H^*(f; \mathbb{Q})$ is an isomorphism.*

Conjecture 2.32 can be adapted to rational homotopy theory as follows (see e.g., Félix, Halperin, and Thomas [FHT01, Section 39, Problem 8]):

Conjecture 2.47 Let M and N be two simply connected closed manifolds. If $M \simeq_{\mathbb{Q}} N$ then $\mathsf{Conf}_M(r) \simeq_{\mathbb{Q}} \mathsf{Conf}_N(r)$ for all $r \geq 0$.

Remark 2.48 Even if Conjecture 2.32 turns out to be true, this would not automatically settle Conjecture 2.47. Indeed, while the conclusion of the second conjecture $(\mathsf{Conf}_M(r) \simeq_{\mathbb{Q}} \mathsf{Conf}_N(r))$ is weaker, so is its hypothesis $(M \simeq_{\mathbb{Q}} N)$.

Let us now introduce the algebraic part of rational homotopy theory, which will allow us to refine the above conjecture. The founding idea of Sullivan [Sul77] is that the rational homotopy type of a (simply connected) space is encoded by a purely algebraic data, namely, a commutative differential-graded algebra.

Remark 2.49 There is a theory due to Quillen [Qui69] which encodes rational homotopy types via differential-graded Lie algebras (see Definition 2.77). This theory is, in a sense, dual—as in Koszul duality or Eckmann–Hilton duality—to Sullivan's theory.

Definition 2.50 A *commutative differential graded algebra* (CDGA) is a cochain complex $A = \{A^n\}_{n \geq 0}$ with a differential $d : A^n \to A^{n+1}$, a unit $1 \in A^0$ and a product $A \to A$ which is unitary, associative and graded-commutative ($ba = (-1)^{|a| \cdot |b|} ab$) and satisfies the Leibniz rule ($d(ab) = (da)b + (-1)^{|a|} a(db)$).

Example 2.51 A commutative algebra is a CDGA concentrated in degree zero. The cohomology of a topological space is a CDGA whose differential is zero.

Example 2.52 The dg-algebra of cochains $C^*(X)$ on a space X is not a CDGA as it fails to be graded commutative. It is only graded-commutative up to homotopy.

The construction that we introduce below fixes this is issue, at the cost of needing to work over a field of characteristic zero.

Definition 2.53 If V is a cochain complex, we let $S(V)$ be the free CDGA on V. As a vector space, it is isomorphic to the quotient of the tensor algebra $\bigoplus_{k \geq 0} V^{\otimes k}$ by the bilateral ideal generated by $v \otimes w - (-1)^{|v| \cdot |w|} w \otimes v$. The product is given by concatenation of tensors. The differential extends the differential of V through the Leibniz rule.

Example 2.54 If V is a cochain complex concentrated in degree 0 (i.e., just a vector space), then $S(V)$ is the polynomial algebra on V. If V is concentrated in degree 1, then $S(V)$ is the exterior algebra on V. More generally, $S(V)$ is isomorphic, as an algebra, to the tensor product $\mathbb{R}[V^{\text{even}}] \otimes \Lambda(V^{\text{odd}})$ of the polynomial algebra of the even part of V with the exterior algebra of the odd part of V.

Notation 2.55 We will generally indicate the degrees of the generators of a free CDGA by a subscript. For example, $S(x_2, y_3)$ is the free CDGA on two variables of respective degrees 2 and 3. It has a linear basis formed by the monomials of the form $x^k y^l$ where $k \in \mathbb{N}$ and $l \in \{0, 1\}$.

Definition 2.56 A *quasi-isomorphism* of CDGAs is a CDGA morphism that induces an isomorphism in cohomology. If two CDGAs A and B are connected through a zigzag of quasi-isomorphisms, we say that they are *quasi-isomorphic* and we write $A \simeq B$.

We will now define the CDGA of piecewise polynomial forms on a topological space. The philosophy behind the definition is similar to the philosophy behind the definition of singular cohomology: we start by first defining polynomial forms on the standard simplex Δ^n (cf. Fig. 2.4), then we define a form on a space X by considering all the ways to map simplices in X.

Definition 2.57 The *standard simplex* Δ^n is the topological space given by:

$$\Delta^n := \{(t_0, \ldots, t_n) \in \mathbb{R}^{n+1} \mid \forall i, t_i \geq 0; \ t_0 + \cdots + t_n = 1\}. \tag{2.22}$$

The relationships between the various ways of including simplices as faces of others and collapsing simplices onto others are given by the following maps, called cofaces and codegeneracies (where $0 \leq i, j \leq n$):

$$\delta^i : \Delta^{n-1} \to \Delta^n$$

$$(t_0, \ldots, t_{n-1}) \mapsto (t_0, \ldots, t_i, 0, t_{i+1}, \ldots, t_{n-1}), \tag{2.23}$$

$$\sigma^j : \Delta^{n+1} \to \Delta^n$$

$$(t_0, \ldots, t_{n+1}) \mapsto (t_0, \ldots, t_{j-1}, t_j + t_{j+1}, t_{j+2}, \ldots, t_{n+1}). \tag{2.24}$$

Fig. 2.4 The first three standard simplexes $\Delta^n \subset \mathbb{R}^{n+1}$

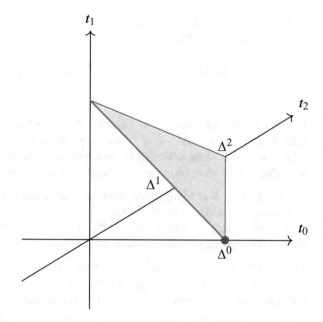

Definition 2.58 The *algebra of polynomial forms on* Δ^n, denoted by Ω_n, is the quotient of the free graded-commutative algebra generated by the symbols $t_0, \ldots, t_n, dt_0, \ldots, dt_n$, where $\deg t_i = 0$, $\deg dt_i = 1$, by the ideal generated by the relations $t_0 + \cdots + t_n = 1$ and $dt_0 + \cdots + dt_n = 0$. The differential is given on the generators by $d(t_i) = dt_i$ and $d(dt_i) = 0$.

The collection $\Omega_\bullet = \{\Omega_n\}_{n \geq 0}$ is equipped with operations that are dual to the operations δ^i and σ^j above:

$$d_i : \Omega_n \to \Omega_{n-1} \ (0 \leq i \leq n) \qquad s_j : \Omega_n \to \Omega_{n+1} \ (0 \leq i \leq n)$$

$$t_k \mapsto \begin{cases} t_k, & \text{if } k < i, \\ 0, & \text{if } k = i, \\ t_{k-1}, & \text{if } k > i; \end{cases} \qquad t_k \mapsto \begin{cases} t_k, & \text{if } k < i, \\ t_k + t_{k+1}, & \text{if } k = i, \\ t_{k+1}, & \text{if } k > i. \end{cases} \qquad (2.25)$$

Remark 2.59 The operations ∂^i and σ^j satisfy a number of relations that make the collection $\{\Delta^n\}_{n \geq 0}$ into what is called a cosimplicial space. Similarly, the operations d_i and s_j satisfy relations that makes $\{\Omega_n\}_{n \geq 0}$ into a simplicial CDGA. We refer to e.g. Goerss and Jardine [GJ99] for a more in-depth treatment.

Definition 2.60 Let X be any space. The CDGA of *piecewise polynomial forms* on X is denoted by $\Omega_{PL}^*(X)$. It is given in degree k by

$$\Omega_{PL}^k(X) = \left\{ \omega = (\omega_f \in \Omega_n^k)_{f:\Delta^n \to X} \mid d_i(\omega_f) = \omega_{f \circ \delta^i}, \; s_j(\omega_j) = \omega_{f \circ \sigma^j} \right\}. \tag{2.26}$$

In other words, an elements of $\Omega_{PL}^k(X)$ is a collection $\omega = (\omega_f)$ indexed by continuous maps $f : \Delta^n \to X$ of forms on the appropriate simplex of degree k, subject to the relations indicated in the definition. The differential and the product are defined term-wise, i.e., $(\omega\omega')_f = \omega_f \omega'_f$ and $(d\omega)_f = d(\omega_f)$.

Remark 2.61 If X is triangulated, then one can define a quasi-isomorphic CDGA in a simpler way. An element of degree k is given by a form of degree k on each simplex of X, subject to the condition that if two simplices meet along a face, then the two corresponding forms coincide on that face.

Theorem 2.62 (Sullivan [Sul77]) *Let X be a simply connected topological space of finite type. Then $\Omega_{PL}^*(X)$ is quasi-isomorphic, as a (non-commutative) DGA, to the algebra of the singular cochains of X with rational coefficients $C^*(X; \mathbb{Q})$.*

As a corollary, the cohomology of the CDGA $\Omega_{PL}^*(X)$ is isomorphic to the rational cohomology of X. The quasi-isomorphism is easy to write down. Given a k-form $\omega = (\omega_f) \in \Omega_{PL}^*(X)$, we define a k-cochain $\int \omega \in C^k(X)$ as follows. For any k-simplex $f : \Delta^k \to X$, which represents a generator of $C_k(X)$, write $\omega_f = P(t_1, \ldots, t_k)dt_1 \ldots dt_k$ where $P(t_1, \ldots, t_k)$ is some polynomial. Then $\langle \int \omega, f \rangle$ is simply equal to the integral of P on the standard k-simplex.

The following theorem is at the heart of Sullivan's rational homotopy theory.

Remark 2.63 In that theorem, we use the notion of a category C localized at a class of morphisms $\mathcal{W}$, generally denoted $C[W^{-1}]$. We refer to Gabriel and Zisman [GZ67] for the precise definition. Briefly, the objects of $C[\mathcal{W}]$ are the objects of C, and the set of morphisms $\mathrm{Hom}_{C[W^{-1}]}(X, Y)$ between two objects are the zigzags of morphisms of the following form:

$$X \xleftarrow[\sim]{f_1} X_1 \xrightarrow{f_2} X_2 \xleftarrow[\sim]{f_3} X_3 \xrightarrow{f_4} \ldots \xleftarrow[\sim]{f_n} X_n \xrightarrow{f_{n+1}} Y, \tag{2.27}$$

where the morphisms that go from right to left (indicated by a $\sim$) are assumed to belong to $\mathcal{W}$. Composition is simply concatenation of zigzags. Furthermore, zigzags are modded out by the relation generated by two kinds of identifications: if two consecutive arrows go in the same direction ($A \xrightarrow{f} B \xrightarrow{g} C$ or the reverse), then they are identified with their composition ($A \xrightarrow{g \circ f} C$ or the reverse); and if two consecutive arrows are given by the same morphism in opposite directions ($A \xleftarrow[\sim]{f} B \xrightarrow[\sim]{f} A$ or the reverse), then they are removed from the zigzag and replaced with the object at the extremities (i.e., A).

Note that there are size issues in this definition. Even if C is locally small (i.e., morphisms between two fixed objects form a set), then $C[W^{-1}]$ may not necessarily be locally small. Quillen's theory of model categories [Qui67] provides a way to construct $C[W^{-1}]$ in a way that solves these issues, provided that C is endowed with auxiliary data (a class of fibrations and a class of cofibrations) that satisfy restrictive conditions. We refer to e.g. [DS95] for an introduction of the subject.

Theorem 2.64 (Sullivan [Sul77]) *The functor* Ω^*_{PL} *induces an equivalence between the following categories:*

- *the category of simply connected topological spaces of finite type, localized at rational equivalences;*
- *the category of finite type CDGAs A such that $A^0 = \mathbb{Q}$ and $A^1 = 0$, localized at quasi-isomorphisms.*

Remark 2.65 To prove this theorem, it is necessary to go through the category of 1-reduced simplicial sets. This is notably due to the fact that even if a space X is simply connected, then one does not necessarily have $\Omega^0_{PL}(X) = \mathbb{Q}$ and $\Omega^1_{PL}(X) = 0$.

Corollary 2.66 *Let X and Y be two simply connected topological spaces of finite type. Then*

$$X \simeq_{\mathbb{Q}} Y \iff \Omega^*_{PL}(X) \simeq \Omega^*_{PL}(Y). \tag{2.28}$$

Definition 2.67 Let X be a topological space. A *model* (or *rational model* if one wants to emphasize the field of coefficients) of X is CDGA A which is quasi-isomorphic to the CDGA $\Omega^*_{PL}(X)$.

Thanks to Sullivan's theorem, a model of X "knows" completely the rational homotopy type of X. In particular, if two spaces have the same model, then they are rationally equivalent.

It is possible to make many computations using a model A of X. First of all, rational cohomology can be computed from model thanks to Theorem 2.62.

Corollary 2.68 *Let A be a model of a space X. Then there is an isomorphism of commutative-graded algebras $H^*(X; \mathbb{Q}) \cong H^*(A)$.*

Moreover, the Massey products (see Definition 2.30) of X can be explicitly computed from the model A using a technique known as the homotopy transfer theorem. We refer to e.g. [LV12, Section 9.4.5].

Second of all, rational homotopy groups can be computed from a model of X. Before stating the result, we first introduce the notion of minimal model.

Definition 2.69 A *quasi-free* CDGA is a CDGA of the form $A = (S(V), d)$ which is free as a graded commutative algebra, where V is a graded vector space (but the differential is not necessarily induced by a differential on V).

Remark 2.70 Let V be a graded vector space. The differential of a quasi-free CDGA of the form $A = (S(V), d)$ is completely determined by its value on the generators, $d : V \to S(V)$, thanks to the Leibniz rule.

Definition 2.71 A *minimal CDGA* is a quasi-free CDGA $A = (S(V), d)$ such that $V = V^{\geq 2}$ and the differential of a generator is at least quadratic, i.e., $d(V) \subset S^{\geq 2}(V)$.

Example 2.72 The CDGA $(S(x_{2k}, y_{4k-1}), dy = x^2)$ is minimal (where x is some variable of degree $2k$ and y some variable of degree $4k - 1$). The CDGA $(S(x_{2k}, z_{2k-1}), dz = x)$ is not minimal, however.

The main interest of minimal model comes from the following results:

Proposition 2.73 *Two minimal CDGAs are quasi-isomorphic if and only if they are isomorphic. Any simply connected CDGA of finite type is quasi-isomorphic to a minimal CDGA.*

Proof *(Idea)* The proof of the first result is in two steps:

1. Any quasi-isomorphism $f : A \to (S(V), d)$ whose target is a minimal CDGA admits a right inverse up to homotopy. This right inverse is built by induction on the degree of the generators. This induction is possible because the minimality condition implies that $d(V^n) \subset S^{\geq 2}(V^{<n})$, i.e., the differential of a generator of degree n is a sum of nontrivial products of generators that all have degree less than n. It is thus possible to start from the lowest degree, and for each induction step, only generators that have already been considered will appear in the differential.
2. We can thus assume that were are given a direct quasi-isomorphism $f :$ $(S(V), d) \to (S(V'), d')$ between two minimal CDGAs. One then checks easily (using a similar argument as the previous one) that f must be an isomorphism on generators.

For the second result, the minimal model is also built by induction on degree. One first starts by adding a generator for every class in a basis of $H^2(A)$. These representatives can satisfy relations in $H^*(A)$, so one new generators (which must be of degree at least 3) whose differentials kill the representatives of these relations. One thus obtain a minimal CDGA and a morphism into A which is a quasi-isomorphism up to degree 2. Then one starts the procedure again, adding generators of degree 3 and killing relations (by adding generators of degree at least $2 + 3 = 5$), so that the second cohomology group is left unchanged but the third one becomes $H^3(A)$. The procedure then continues inductively and the colimit has the correct homotopy type.   □

As a consequence of the previous proposition, any space X admits a unique (up to isomorphism) model which is minimal as a CDGA.

Definition 2.74 Let X be a simply connected space of finite type. The *minimal model* of X is the model of X which is minimal as a CDGA.

Theorem 2.75 (See e.g. [FHT01, Theorem 15.11]) *Let X be a simply connected space of finite type and let $(S(V), d)$ be its minimal model. Then for all $n \geq 2$, there is an isomorphism, natural in X:*

$$\varphi_n : V^n \xrightarrow{\cong} \mathrm{Hom}(\pi_n(X), \mathbb{Q}). \tag{2.29}$$

The knowledge of a model of a space allows one to compute its minimal model (essentially by induction) and thus to compute the rational homotopy groups of that space. There is moreover a natural operation on homotopy groups called the Whitehead product (or bracket). It is defined as follows. For integers $k, l \geq 1$, the product $\mathbb{S}^k \times \mathbb{S}^l$ can be obtained from the wedge sum $\mathbb{S}^k \vee \mathbb{S}^l$ by gluing a $(k+l)$-cell along an attaching map $f_{k,l} : \mathbb{S}^{k+l-1} \to \mathbb{S}^k \vee \mathbb{S}^l$.

Definition 2.76 Let X be a space and $\alpha \in \pi_k(X)$, $\beta \in \pi_l(X)$. The *Whitehead product* [Whi41] is the element $[\alpha, \beta] \in \pi_{k+l-1}(X)$ represented by:

$$[\alpha, \beta] : \mathbb{S}^{k+l-1} \xrightarrow{f_{k,l}} \mathbb{S}^k \vee \mathbb{S}^l \xrightarrow{\alpha \vee \beta} X. \tag{2.30}$$

The Whitehead product makes the shifted space $\bigoplus_{k \geq 0} \pi_{k+1}(X)$ into a Lie algebra. Let us define this notion in the case of rational coefficients:

Definition 2.77 A *Lie algebra* is a graded vector space $\mathfrak{g}$ equipped with a bilinear map $[-, -] : \mathfrak{g} \otimes \mathfrak{g} \to \mathfrak{g}$ which is antisymmetric: for all $x, y \in \mathfrak{g}$,

$$\forall x, y \in \mathfrak{g}, \ [y, x] = -(-1)^{|x| \cdot |y|}[x, y], \tag{2.31}$$

and which satisfies the Jacobi relation: for all $x, y, z \in \mathfrak{g}$,

$$(-1)^{|x| \cdot |z|}[x, [y, z]] + (-1)^{|x| \cdot |y|}[y, [z, x]] + (-1)^{|y| \cdot |z|}[z, [x, y]] = 0. \tag{2.32}$$

Remark 2.78 With this definition, one can only prove that $2[x, x] = 0$ if $x \in \mathfrak{g}$ has even degree, and that $3[[y, y], y] = 0$ if $y \in \mathfrak{g}$ has odd degree. Over a field of positive characteristic, or over $\mathbb{Z}$, one must add the conditions that $[x, x] = 0$ and $[[y, y], y] = 0$ for any x of even degree and y of odd degree to get the correct definition.

Proposition 2.79 *Let X be a simply connected space of finite type and $(S(V), d)$ its minimal model. Let $d_2 : V \to V^{\otimes 2}/\mathfrak{S}_2$ be the quadratic part of d, i.e., the projection of the restriction of d to the generators onto the weight 2 component of $S(V)$. Then d_2 is dual to the Whitehead product under the isomorphism of Theorem 2.75:*

$$\forall v \in V, \ \forall \alpha, \beta \in \pi_{*+1}(X), \ \langle \varphi(v), [\alpha, \beta] \rangle = \langle (\varphi \otimes \varphi)(d_2(v)), \alpha \otimes \beta \rangle. \tag{2.33}$$

Let us now give an example of a model.

Example 2.80 Let us consider the sphere $\mathbb{S}^n$ of dimension n. It can be proved quite easily that a model of $\mathbb{S}^n$ is given by its cohomology, $H^*(\mathbb{S}^n) = \mathbb{Q}1 \oplus \mathbb{Q}v$, where $\deg v = n$ and $v^2 = 0$. There are two cases:

- Suppose that n is odd. Then $H^*(\mathbb{S}^n)$ is actually the free CDGA on one variable of degree n (as a variable of odd degree anti-commutes with itself and thus automatically squares to zero). Choose any closed representative $\mathrm{vol}_{\mathbb{S}^n} \in \Omega^n_{\mathrm{PL}}(\mathbb{S}^n)$ of the volume form of $\mathbb{S}^n$. Then there is a well-defined morphism $H^*(\mathbb{S}^n) \to \Omega^*_{\mathrm{PL}}(\mathbb{S}^n)$ which maps v to $\mathrm{vol}_{\mathbb{S}^n}$. This morphism is obviously surjective in cohomology, and since both CDGAs have the same cohomology, the claim follows. Note that $H^*(\mathbb{S}^n)$ is minimal in that case, so it is the minimal model of $\mathbb{S}^n$.

- Suppose now that n is even. Choose as before any closed representative $\mathrm{vol}_{\mathbb{S}^n} \in \Omega^n_{\mathrm{PL}}(\mathbb{S}^n)$ of the volume form of $\mathbb{S}^n$. The issue is that now, the square of $\mathrm{vol}_{\mathbb{S}^n}$ may be nonzero. However, thanks to the computation of the cohomology of the sphere, $\mathrm{vol}^2_{\mathbb{S}^n}$ must be a coboundary, say $\mathrm{vol}^2_{\mathbb{S}^n} = d\alpha$. Then we can build a zigzag of quasi-isomorphisms of CDGAs:

$$H^*(\mathbb{S}^n) \longleftarrow \left(S(x_n, y_{2n-1}), dy = x^2\right) \longrightarrow \Omega^*_{\mathrm{PL}}(\mathbb{S}^n)$$
$$v \longleftarrow\!\!\mid x \mid\!\longrightarrow \mathrm{vol}_{\mathbb{S}^n}$$
$$0 \longleftarrow\!\!\mid y \mid\!\longrightarrow \alpha \qquad\qquad (2.34)$$

The CDGA in the middle is freely generated (as a CGA) by two variables x (of degree n) and y (of degree $2n - 1$). The differential is the unique derivation that satisfies $dx = 0$ and $dy = x^2$. It can be seen easily that the left-pointing map is a quasi-isomorphism. Therefore, since the right-pointing map is clearly surjective in cohomology, the claim follows. Note that the CDGA in the middle is the minimal model of $\mathbb{S}^n$.

The sphere is thus an example of formal space (see Sect. 2.4). We can also recover a theorem of Serre:

- if n is odd then $\pi_n(\mathbb{S}^n) \otimes_{\mathbb{Z}} \mathbb{Q} = \mathbb{Q}$ and all other homotopy groups are torsion;
- if n is even then $\pi_n(\mathbb{S}^n) \otimes_{\mathbb{Z}} \mathbb{Q} = \pi_{2n-1}(\mathbb{S}^n) \otimes_{\mathbb{Z}} \mathbb{Q} = \mathbb{Q}$ and all other homotopy groups are torsion; moreover the Whitehead bracket of the degree n with itself is equal to the degree $2n - 1$ generator.

Returning once again to the configuration spaces, we arrive at a finer version of Conjecture 2.47:

Conjecture 2.81 Let M be a simply connected closed manifold. It is possible to find an explicit model of $\mathrm{Conf}_M(r)$ that depends only on a model of M.

If this conjecture holds, then Conjecture 2.47 also holds thanks to Sullivan's theorem.

In these notes, we will actually focus on real homotopy types, not rational homotopy types. While real homotopy types are slightly weaker, they are still sufficient for most computations.

Definition 2.82 Two simply connected spaces of finite type X and Y have the same *real homotopy type* if $\Omega^*_{\text{PL}}(X) \otimes_{\mathbb{Q}} \mathbb{R}$ and $\Omega^*_{\text{PL}}(Y) \otimes_{\mathbb{Q}} \mathbb{R}$ are quasi-isomorphic as CDGAs. A *real model* of such a space X is a CDGA (with real coefficients) which is quasi-isomorphic to $\Omega^*_{\text{PL}}(X) \otimes_{\mathbb{Q}} \mathbb{R}$.

Remark 2.83 Unlike rational equivalences, real equivalences cannot always be realized by zigzags of continuous maps that induces isomorphisms on real cohomology.

Example 2.84 ([FOT08, Example 2.38]) Let $\alpha \in \mathbb{Q}$ be a positive rational parameter. We define a CDGA

$$A_\alpha = \big(S(e_2, x_4, y_7, z_9), d_\alpha\big), \tag{2.35}$$

where the indices denote the degree, together with the differential:

$$d_\alpha e = 0, \; d_\alpha x = 0, \; d_\alpha y = x^2 + \alpha e^4, \; d_\alpha z = e^5. \tag{2.36}$$

For two different parameters α and α', the CDGAs A_α and $A_{\alpha'}$ are quasi-isomorphic if and only if α/α' is a square. They are thus always quasi-isomorphic over $\mathbb{R}$, but not always over $\mathbb{Q}$.

Example 2.85 ([FOT08, Example 3.7]) The theory is obviously generalizable to $\mathbb{C}$. Consider the complex projective plane $\mathbb{CP}^2$ with its usual orientation and let $\overline{\mathbb{CP}}^2$ be the same manifold with reverse orientation. The connected sums $X = \mathbb{CP}^2 \# \mathbb{CP}^2$ and $Y = \mathbb{CP}^2 \# \overline{\mathbb{CP}}^2$ have the same complex homotopy type, but not the same real homotopy type. One can see (with a proof similar to Example 2.80 in the even case) that a model for X is given by:

$$H^*(X) = S(x, y)/(x^2 - y^2, xy, x^3, y^3), \tag{2.37}$$

where $\deg x = \deg y = 2$, whereas a model for Y is given by:

$$H^*(Y) = S(x, y)/(x^2 + y^2, xy, x^3, y^3). \tag{2.38}$$

Then there is an isomorphism over $\mathbb{C}$ between these two models given by $x \mapsto x$, $y \mapsto iy$. However, these two models are not quasi-isomorphic over $\mathbb{R}$. This can be seen by considering their intersection forms, which are quadratic forms given by

$$\begin{aligned} q_X : H^2(X) &\to \mathbb{R} \\ \alpha &\mapsto \langle \alpha^2, [X] \rangle \end{aligned} \tag{2.39}$$

where $[X] \in H_4(X)$ is an orientation (and similarly for Y). The signature of q_X is $(2, 0)$ while the signature of q_Y is $(1, 1)$.

Let us now focus on configuration spaces once again. Conjecture 2.47 stated that the rational homotopy type of $\mathsf{Conf}_M(r)$ only depends on the rational homotopy type of M. This conjecture can be modified slightly to yield a different one.

Conjecture 2.86 Let M be a simply connected closed manifold. It is possible to find an explicit *real* model of $\mathsf{Conf}_M(r)$ that depends only on a real model of M.

A positive answer to this conjecture would make it possible to make calculations on the real homotopy of $\mathsf{Conf}_M(r)$ (cohomology, homotopy groups, etc.) simply by knowing the homotopy type of M.

2.4 Configuration Spaces of Euclidean Spaces

The basic building bricks manifolds are the Euclidean spaces $\mathbb{R}^n$. Since any manifold M can be obtained by gluing copies of $\mathbb{R}^n$, the configuration space $\mathsf{Conf}_M(r)$ can thus be obtained by gluing copies of $\mathsf{Conf}_{\mathbb{R}^n}(k)$ for $k \leq r$ in a rather complicated way (with points potentially in different charts). To understand the homotopy type of the configuration spaces of M, we will therefore start by looking at those of $\mathbb{R}^n$.

The cohomology of the configuration spaces of $\mathbb{R}^n$ is well known. Let us first introduce the following notation, that will become clear in Chap. 5 (as $\mathsf{Conf}_{\mathbb{R}^n}(r)$ has the cohomology of the little disks operad, E_n).

Definition 2.87 Let $n \geq 2$ and $r \geq 0$ be integers. We let $\mathsf{e}_n^\vee(r)$ be the graded-commutative algebra with the following presentation, where $\deg \omega_{ij} = n - 1$:

$$\mathsf{e}_n^\vee(r) := \frac{S(\omega_{ij})_{1 \leq i \neq j \leq r}}{\left(\omega_{ji} = (-1)^n \omega_{ij},\ \omega_{ij}^2 = 0,\ \omega_{ij}\omega_{jk} + \omega_{jk}\omega_{ki} + \omega_{ki}\omega_{ij} = 0\right)} \tag{2.40}$$

Theorem 2.88 (Arnold [Arn69], Cohen [Coh76]) *The cohomology of* $\mathsf{Conf}_{\mathbb{R}^n}(r)$ *is isomorphic to* $\mathsf{e}_n^\vee(r)$ *as a graded-commutative algebra.*

Intuitively, the class ω_{ij} counts how many times the points indexed by i and j revolve around each other. More formally, recall from Lemma 2.21 that $\mathsf{Conf}_{\mathbb{R}^n}(2)$ is homotopy equivalent to a sphere through the map $\mathsf{Conf}_{\mathbb{R}^n}(2) \to \mathbb{S}^{n-1}$, $(x, y) \mapsto (x - y)/\|x - y\|$. Let $\theta_{ij} : \mathsf{Conf}_{\mathbb{R}^n}(r) \to \mathsf{Conf}_{\mathbb{R}^n}(2) \to \mathbb{S}^{n-1}$ be the map given by

$$\theta_{ij}(x_1, \ldots, x_n) := \frac{x_i - x_j}{\|x_i - x_j\|}. \tag{2.41}$$

The class ω_{ij} is the pullback of the volume form of the sphere along θ_{ij}. The relations can be interpreted as follows:

- the relation $\omega_{ij} = (-1)^n \omega_{ji}$ says that reversing the orientation of the sphere can introduce a sign;

Fig. 2.5 The element
$\Gamma = \omega_{12}\omega_{24} + \frac{1}{2}\omega_{23} \in e_n^{\vee}(4)$
seen as a linear combination
of graphs

$$\Gamma = \left(\begin{matrix} \text{3} & \text{4} \\ & | \\ \text{1} - \text{2} \end{matrix} \right) + \frac{1}{2} \left(\begin{matrix} \text{3} & \text{4} \\ & \times \\ \text{1} & \text{2} \end{matrix} \right)$$

- the relation $\omega_{ij}^2 = 0$ says that the volume form has vanishing square;
- the relation $\omega_{ij}\omega_{jk} + \omega_{jk}\omega_{ki} + \omega_{ki}\omega_{ij} = 0$ is, in a sense, dual to the Jacobi relation which describes what happens when three points interact by rotating around common centers (for example points 1, 2 and 4 in Fig. 2.7)

Let us now give a graphical reinterpretation of Eq. (2.40) that will be useful throughout this volume. The elements of $e_n^{\vee}(r)$ can be seen as formal linear combinations of graphs with r numbered vertices. Such a graph, with edges $(i_1, j_1), \ldots, (i_l, j_l)$, corresponds to the word $\omega_{i_1 j_1} \ldots \omega_{i_l j_l}$. See Fig. 2.5 for example. Note that since the generators are required to square to zero, there can be no double edge in a graph. Moreover, there are no generators of the type ω_{ii}, so there are no tadpoles, i.e., edges between a vertex and itself.

Moreover, each graph is equipped with some orientation data:

- if n is even, then the set of edges is ordered;
- if n is odd then the edges are oriented.

A change of order, resp. a change of orientation, induces a change of sign. This corresponds either to the fact that the generators anti-commute (if n is even) or that exchanging the roles of i and j in ω_{ij} introduces a sign. Note that we will typically draw pictures of graphs without the orientation data (as in e.g., Fig. 2.5). Orienting or ordering edges is necessary to get precise signs out of these pictures.

The space of all linear combinations of such graphs is modded out by local relations involving three vertices. These local relations correspond to the three-term Arnold relations, $\omega_{ij}\omega_{jk} + \omega_{jk}\omega_{ki} + \omega_{ki}\omega_{ij} = 0$. These local relations can be drawn as follows, where we are looking at a linear combination of three graphs which are identical outside of the dashed circle (with appropriate signs depending on how edges are oriented or ordered):

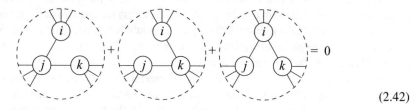

$$\tag{2.42}$$

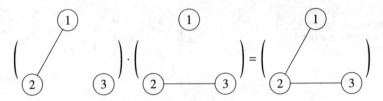

Fig. 2.6 The product $\omega_{12} \cdot \omega_{23} \in e_n^\vee(3)$ interpreted graphically

Finally, the product of the CDGA consists in taking the union of graphs, or in other words gluing them together along their vertices (with matching indices). See Fig. 2.6 for an example.

The following lemma is well-known. It dates back to Arnold [Arn69] for the case $n = 2$, and Cohen [Coh73a] for $n \geq 3$. We will not give the original proof (which uses an inductive argument on the number of points in a configuration). We will instead use the proof as an excuse to introduce the theory of Poincaré–Birkhoff–Witt bases, an important combinatorial tool. A reader interested only in the topology of configuration spaces can skip this proof.

Lemma 2.89 *Let $r \geq 0$ be an integer. A basis for $e_n^\vee(r)$ is given by monomials of the type:*

$$\omega_{i_1 j_1} \dots \omega_{i_k j_k}, \tag{2.43}$$

where $j_1 < \dots < j_k$ and $i_l < j_l$ for all l.

Proof It is easy to see that these monomials span the algebra using the previous graphical description. These monomials correspond exactly to graphs where the edges are ordered by their targets, the edges are all oriented from lower vertices to higher ones, and no two edges have the same target. Using the relations $\omega_{ij}^2 = 0$, $\omega_{ji} = (-1)^n \omega_{ij}$, and the graded commutativity of the algebra, the first two conditions can be achieved. Moreover, any graph such that two edges have the same target can be rewritten in terms of graphs where edges have different targets using the 3-term Arnold relation of Eq. (2.42).

It is harder to show that these monomials are linearly independent. There are several approaches. One is to use the graphs-trees pairing that we use in the proof of Theorem 2.88. Another is to use the theory of commutative Poincaré–Birkhoff–Witt bases, as we now explain very briefly. We refer to e.g., Polishchuk and Positselski [PP05, Section 4.8] for the general theory.

Thanks to the relation $\omega_{ji} = (-1)^n \omega_{ij}$, we can take as generators the elements ω_{ij} for $i < j$. We then work in the free graded-commutative algebra generated by these elements. We decide to order these generators by the lexicographic order, i.e.:

$$\omega_{ij} < \omega_{kl} \iff i < k \text{ or } (i = k \text{ and } j < l). \tag{2.44}$$

We then order monomials lexicographically. Then the remaining two kinds of relations are quadratic and can be seen as so-called "rewriting rules", where a higher monomial (the "dominant term") is seen as rewritten in terms of lower monomials:

$$\omega_{ij}\omega_{ij} \rightsquigarrow 0, \qquad\qquad\qquad \text{for } i < j; \qquad (2.45)$$

$$\omega_{ik}\omega_{jk} \rightsquigarrow \omega_{ij}\omega_{jk} - \omega_{ij}\omega_{ik} \qquad \text{for } i < j < k. \qquad (2.46)$$

The first relation simply means that graphs containing double edges are rewritten to zero. The second relation means that if a graph contains two edges with the same target, then that graph is rewritten using Eq. (2.42). The claimed basis is precisely given by the monomials that cannot be rewritten in this way.

The general theory gives us a criterion to check whether these monomials are linearly independent. A monomial is called "critical" if is minimal among monomials containing two overlapping dominant terms of rewriting rules. For example, the monomial $\omega_{ik}\omega_{jk}\omega_{jk}$ is critical, as it contains the two dominant terms $\omega_{ik}\omega_{jk}$ and $\omega_{jk}\omega_{jk}$, which overlap at the middle term. The monomials $\omega_{ij}\omega_{jk}\omega_{jk}$ and ω_{ij}^4 are not critical: the first because it does not contain two overlapping dominant terms of rewriting rules, the second because it is not minimal (it contains the critical monomial ω_{ij}^3).

The criterion mentioned above says that if these critical monomials are confluent, i.e., if by applying either one of the two overlapping rewriting rules and then applying rewriting rules again, we can get to a common monomial. In our case, since the rewriting rules are all quadratic, then critical monomials must be cubic. The critical monomials are of three kinds, and we can check easily by hand that they are confluent.

1. The critical monomial $(\omega_{ij}^2)\omega_{ij} = \omega_{ij}(\omega_{ij}^2)$ is clearly confluent, as applying the rewriting rule to either factor yields zero.
2. The critical monomial $\omega_{ik}(\omega_{jk}\omega_{jk}) = (\omega_{ik}\omega_{jk})\omega_{jk}$ (where $i < j < k$) is confluent:

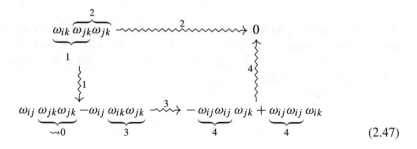

$$(2.47)$$

3. The critical monomial $(\omega_{il}\omega_{jl})\omega_{kl} = \omega_{il}(\omega_{jl}\omega_{kl})$ (where $i < j < k < l$) is also confluent. It is a straightforward (if long) exercise to check this fact. It is helpful to draw monomials as graphs; the critical monomial is a graph containing three edges with the same target. $\qquad\qquad\qquad\qquad\qquad\qquad\qquad\qquad\qquad\qquad\qquad\qquad\qquad \square$

This lemma allows us to compute the dimension of $e_n^\vee(r)$, simply by counting the degrees of the monomials.

Corollary 2.90 *The Poincaré polynomial of* $e_n^\vee(r)$ *is given by:*

$$\sum_{d \in \mathbb{N}} \dim(e_n^\vee(r)^d) \cdot t^d = \prod_{i=0}^{r-1}(1 + it^{n-1}) \in \mathbb{N}[t]. \tag{2.48}$$

Proof Thanks to Lemma 2.89, we see that $e_n^\vee(r)$ is isomorphic to the tensor product $V_1 \otimes \cdots \otimes V_r$, where V_j is the graded vector space spanned by $1, \omega_{1j}, \ldots, \omega_{(j-1)j}$. The Poincaré polynomial of V_j is thus $1 + (j-1)t^{n-1}$ and by multiplicativity of the Poincaré polynomial, the corollary follows.   □

Proof (Sketch of Proof of Theorem 2.88) We refer to the presentation given by Sinha [Sin13] for what follows. Let us consider the projection:

$$p_{1,\ldots,r-1} : \mathsf{Conf}_{\mathbb{R}^n}(r) \to \mathsf{Conf}_{\mathbb{R}^n}(r-1) \tag{2.49}$$

that forgets the last point of a configuration. This map is a locally trivial fiber bundle thanks to [FN62a, Theorem 3]. Its fiber is $\mathbb{R}^n \setminus \{x_1^0, \ldots, x_{r-1}^0\}$, where $x^0 \in \mathsf{Conf}_{\mathbb{R}^n}(r-1)$ is any base point. This fiber has the homotopy type of the wedge of spheres $\bigvee_{i=1}^{r-1} \mathbb{S}^{n-1}$. Thanks to the Serre spectral sequence (see e.g., Bott and Tu [BT82. §15]), the Betti numbers of $\mathsf{Conf}_{\mathbb{R}^n}(r)$ are bounded above by the Betti numbers of the tensor product:

$$H^*(\mathsf{Conf}_{\mathbb{R}^n}(r-1)) \otimes H^*\left(\bigvee_{i=1}^{r-1} \mathbb{S}^{n-1}\right)$$

$$\cong H^*(\mathsf{Conf}_{\mathbb{R}^n}(r-1)) \oplus \bigoplus_{i=1}^{r-1} H^{*-n+1}(\mathsf{Conf}_{\mathbb{R}^n}(r-1)). \tag{2.50}$$

We can thus show by induction that the Betti numbers of $\mathsf{Conf}_{\mathbb{R}^n}(r)$ are at most equal to the dimensions of the components of $e_n^\vee(r)$ (which were computed in Corollary 2.90). Moreover, if equality is reached, then the cohomology group considered is free as an abelian group.

Let us consider the following map:

$$\varphi : e_n^\vee(r) \to H^*(\mathsf{Conf}_{\mathbb{R}^n}(r)) \tag{2.51}$$

$$\omega_{ij} \mapsto \theta_{ij}^*(\mathrm{vol}_{\mathbb{S}^{n-1}}). \tag{2.52}$$

We can check easily that the relations between the generators are satisfied. Therefore φ defines a morphism of commutative graded algebras. Using the result on Betti numbers, it is enough to show that φ is injective in order to prove the result.

We now use the graphical interpretation of $e_n^\vee(r)$ in order to prove the result. For each graph, a matching homology class can be constructed. The method uses the "solar systems" construction due to Cohen [Coh73a]. We refer to [Sin13, Proposition 2.2] for a detailed treatment. Let us now quickly sketch it.

We will first review some definitions about trees. Informally, we will call "rooted planar binary tree" a finite graph without cycles whose vertices are either trivalent or univalent, together with a distinguished univalent vertex called the "root". The other univalent vertices will be called the "leaves" of the tree. The "height" $h(v)$ of a vertex v will be the length of the unique path between v and the root. The "parent" $p(v)$ of that vertex (if it exists) is the vertex which comes immediately after v on that same path. More formally, we define the set $\mathcal{T}(h)$ of rooted planar binary trees of height at most h in the following inductive manner:

$$\mathcal{T}(0) := \{*\} \quad \text{(the unique tree of height 0)}, \tag{2.53}$$

$$\mathcal{T}(h+1) := \big\{(T_l, T_r) \mid T_l, T_r \in \bigcup_{k=0}^{h} \mathcal{T}(k)\big\}, \tag{2.54}$$

$$\mathcal{T} := \bigcup_{n \geq 0} \mathcal{T}(h). \tag{2.55}$$

Moreover, a forest is a disjoint union of such trees, forming the set:

$$\mathcal{F} := \coprod_{d \geq 0} \mathcal{T}^d.$$

For example, the forest which appears in Fig. 2.7 (without considering the indices of the leaves) is simply:

$$T = \Big((*, (*, *)), \ (*, *) \Big). \tag{2.56}$$

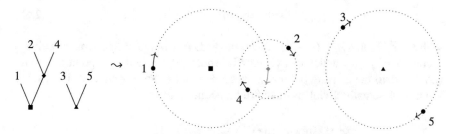

Fig. 2.7 Solar systems: a tree with 5 leaves and 3 internal vertices induces a homology class in $H_{3(2-1)}(\mathrm{Conf}_{\mathbb{R}^2}(5))$. The left system is centered at $(1, 0)$ and the right system is centered at $(2, 0)$

Let $T \in \mathcal{F}$ be a forest equipped with a bijection between its leaves and $\{1, \ldots, n\}$ for some $n \in \mathbb{N}$. One can build a homology class from T as follows. Let $x = (x_1, \ldots, x_r) \in \mathrm{Conf}_{\mathbb{R}^n}(r)$ be a configuration. We inductively define the "center of mass" $c(x, v)$ with respect to a vertex v of T as follows. If ℓ_i is the leaf of T indexed by $i \in \{1, \ldots, n\}$, then $c(x, \ell_i) = x_i$. If v is a trivalent vertex of T, and v' and v'' are its two children, then we define $c(x, v) = (c(x, v') + c(x, v''))/2$. Let $(u_1, \ldots, u_k)$ be the roots of the trees in T. For any $i \in \{1, \ldots, r\}$, there is a unique path $(u_{\rho(i)}, v_1(i), \ldots, v_{h(i)}(i) = i)$ from a root to the leaf ℓ_i. Then there is a map, where $V(T)$ is the set of internal vertices of T and $0 < \tau < 1/3$ is an arbitrary radius:

$$P_T : (\mathbb{S}^{n-1})^{V(T)} \to \mathrm{Conf}_{\mathbb{R}^n}(r)$$

$$(\xi_v)_{v \in V(T)} \mapsto (\rho(i), 0, \ldots, 0) + \sum_{j=1}^{h(i)} \pm \tau^j \xi_{v_j(i)}, \tag{2.57}$$

where the sign $\pm$ is $+1$ if $v_j(i)$ is the left child of its parent, and -1 otherwise. Concretely, we can view the image of P_T as several orbital systems such that the roots are fixed at the points $(1, 0, \ldots, 0), \ldots, (k, 0, \ldots, 0)$, and two children of an internal vertex orbit a common center of mass located at their parent (with a radius depending on the height). We refer to Fig. 2.7 for an example.

The fundamental class of the product of spheres (oriented correctly) thus yields a homology class in $\mathrm{Conf}_{\mathbb{R}^n}(r)$ by pushing it forward along P_T. The homology-cohomology pairing induces a pairing between graphs (i.e., the elements of $e_n^\vee(r)$) and rooted binary forests. One can show combinatorially that this pairing is not degenerate [Sin13, Theorem 4.7], from which the result follows. $\square$

Remark 2.91 In the previous proof, we saw appearing a fundamental object of interest for configuration spaces, the Fadell–Neuwirth fibration [FN62a]. In general, if M is an arbitrary manifold, $Q_m \subset M$ is a subset of cardinality m, and Q_{m+r} is a subset of cardinality $m + r$ that contains Q_m, then there is a locally trivial fiber bundle:

$$p_{1,\ldots,r} : \mathrm{Conf}_{M \backslash Q_m}(n) \to \mathrm{Conf}_{M \backslash Q_m}(r) \tag{2.58}$$

with fiber $\mathrm{Conf}_{M \backslash Q_{m+r}}(n - r)$. When $m \geq 1$, this fibration splits: there exists a section σ of $p_{1,\ldots,r}$ which can visually be understood as adding a point at infinity (i.e., close to the missing point q_1). When M is a Lie group and $m = 0$, then the fibration is actually trivial, as there is a homeomorphism:

$$M \times \mathrm{Conf}_{M \backslash \{1\}}(r) \xrightarrow{\cong} \mathrm{Conf}_M(r + 1)$$

$$(x_0, (x_1, \ldots, x_r)) \mapsto (x_0, x_0^{-1} x_1, \ldots, x_0^{-1} x_r). \tag{2.59}$$

In general, though, the fibration is not trivial. There is a tower of fibrations:

$$
\begin{array}{ccc}
M \setminus \{q_1, \ldots, q_r\} & \hookrightarrow & \mathsf{Conf}_M(r+1) \\
 & & \downarrow \\
M \setminus \{q_1, \ldots, q_{r-1}\} & \hookrightarrow & \mathsf{Conf}_M(r) \\
 & & \downarrow \\
 & & \cdots \\
 & & \downarrow \\
M \setminus \{q_1, q_2\} & \hookrightarrow & \mathsf{Conf}_M(3) \\
 & & \downarrow \\
M \setminus \{q_1\} & \hookrightarrow & \mathsf{Conf}_M(2) \\
 & & \downarrow \\
 & & M
\end{array}
\tag{2.60}
$$

which is extremely useful to study configuration spaces by induction. In particular, as one can see from the long exact sequence of a fibration, if M is a simply connected manifold of dimension at least 3, then $\mathsf{Conf}_M(r)$ is simply connected for all r (see [FN62a, Theorem 9]).

In general, the cohomology of a topological space gives only partial information about the space in question. However, for a certain class of spaces, this information is sufficient to recover the full rational homotopy type.

Definition 2.92 A space X is said to be *formal* if $H^*(X; \mathbb{Q})$ is a model of X in the sense of Definition 2.67.

Example 2.93 The spheres are formal, see Example 2.80.

Example 2.94 ([FOT08, Theorem 1.34]) An H-space is a space X equipped with a map $\mu : X \times X \to X$ which is associative and unitary up to homotopy, i.e. $\mu(\mu(-, -), -) \simeq \mu(-, \mu(-, -))$ and there is an element $e \in X$ such as $\mu(-, e) \simeq \mu(e, -) \simeq \mathrm{id}_X$. Such a space is always automatically formal. Indeed, the cohomology of an H-space is an exterior algebra $\Lambda(z_1, \ldots, z_k)$ on several variables of odd degrees. We can thus define a quasi-isomorphism $\Lambda(z_1, \ldots, z_k) \to \Omega^*_{\mathrm{PL}}(X)$ simply by choosing any closed representatives of the z_i.

Example 2.95 ([FOT08, Proposition 2.99]) If M is a $(p-1)$-connected closed manifold of dimension $\dim M \leq 4p - 2$, then M is formal.

Although we saw earlier that two spaces can have the same real homotopy types but different rational homotopy types, this cannot occur when the spaces are formal.

Theorem 2.96 (Consequence of [Sul77, Theorem 12.1]) *Formality over $\mathbb{Q}$ is equivalent to formality over any field of characteristic zero.*

An important class of examples of formal spaces is the following, which illustrates a deep connection between geometry and topology:

Definition 2.97 A *Kähler manifold* is a complex variety with a Hermitian metric h whose associated 2-form $\omega = -\Im h$ is closed.

Theorem 2.98 (Deligne et al. [DGMS75]) *Compact Kähler manifolds are formal. (More generally, spaces satisfying the dd^c lemma, e.g. Moišezon manifolds, are formal).*

Proof (Idea) Let us consider the real-valued de Rham complex $\Omega^*_{\mathrm{dR}}(M)$ on a Kähler manifold M. Note this this CDGA is quasi-isomorphic to $\Omega^*_{\mathrm{PL}}(X) \otimes_{\mathbb{Q}} \mathbb{R}$. Thanks to Theorem 2.96, it suffices to show that $\Omega^*_{\mathrm{dR}}(M) \simeq H^*(M; \mathbb{R})$.

The de Rham complex is equipped with the exterior differential d, which splits as the sum $d = \partial + \bar{\partial}$, where ∂ differentiates with respect to holomorphic coordinates and $\bar{\partial}$ with respect to anti-holomorphic coordinates. There is another differential $d^c = i(\partial - \bar{\partial})$ which preserves real-valued forms. These two differentials satisfy:

$$d^2 = (d^c)^2 = dd^c + d^c d = 0. \tag{2.61}$$

The fact that M is Kähler implies that it satisfies the "dd^c-lemma": if $\alpha \in \Omega^*_{\mathrm{dR}}(M)$ is a form which is closed with respect to both differentials (i.e., $d\alpha = d^c\alpha = 0$) and exact with respect to d (i.e., $\alpha = d\gamma$ for some γ), then $\alpha = dd^c\beta$ for some form β. Given this lemma, proving formality is a simple matter. One has a zigzag of cochain maps:

$$\left(\Omega^*_{\mathrm{dR}}(M),\, d\right) \overset{i}{\longleftarrow} \left(\ker d^c,\, d\right) \overset{p}{\longrightarrow\!\!\!\!\rightarrow} \left(\ker d^c / \operatorname{im} d^c,\, d\right), \tag{2.62}$$

where i is the inclusion of d^c-closed forms and p is the projection onto d^c-cohomology. Then it is a simply exercise to show that i and p are quasi-isomorphisms, and that d induces the zero differential on $\ker d^c / \operatorname{im} d^c$, which is therefore isomorphic to $H^*(M)$:

- i is injective in cohomology: if $y \in \ker d^c$ is such that $y = dx$ for some form x, then by the lemma $y = d(d^c\beta)$ for some β and $d^c\beta$ is d^c-closed.
- i is surjective in cohomology: if x is a closed form, then $d^c x = dd^c\beta$ for some form, thus $x + d\beta$ is d^c-closed and $i[y] = [x]$.
- d vanishes on $\ker d^c / \operatorname{im} d^c$: if $d^c y = 0$, then $dy = dd^c\beta = -d^c d\beta$ and thus $[dy] = 0$ in the quotient.
- p is surjective in cohomology: if $d^c y = 0$, then $dy = dd^c\beta$ and thus $z = y + d^c\beta$ is a d-closed form mapped to the class of y in the quotient.
- p is injective in cohomology: if $y = d^c x$ then $y = dd^c\beta$ and so y is also d-exact in $\ker d^c$.

□

Let us now go focus on the configuration spaces of $\mathbb{R}^n$. We first have the following result in dimension 2, which uses complex analysis. Although the spaces

$Conf_{\mathbb{C}}(r)$ are Kähler manifolds, they are not compact, so Theorem 2.98 does not apply.

Remark 2.99 The compactness assumption is essential in Theorem 2.98. For example, the configuration spaces $Conf_{\Sigma_g}(r)$ of surfaces of genus $g \geq 2$ are open Kähler manifolds which are not formal for $r \geq 2$. Results of Morgan [Mor78] give ways of finding models of complements of divisors with normal crossings in Kähler manifolds, which can be applied to configuration spaces, see Remark 3.19.

Theorem 2.100 (Arnold [Arn69]) *The configuration spaces of* $\mathbb{C} = \mathbb{R}^2$ *are formal.*

Proof Recall the description of the cohomology of $Conf_{\mathbb{C}}(r)$ from Theorem 2.88. There is a direct morphism from cohomology to forms, defined on generators by:

$$\varphi : H^*(Conf_{\mathbb{C}}(r)) \to \Omega_{dR}^*(Conf_{\mathbb{C}}(r); \mathbb{C})$$

$$\omega_{ij} \mapsto d\log(z_i - z_j) = \frac{dz_i - dz_j}{z_i - z_j}. \tag{2.63}$$

The forms in the target are of course closed. One can easily check that $\varphi(\omega_{ij})^2 = 0$ and that $\varphi(\omega_{ji}) = \varphi(\omega_{ij})$. Finally, a small calculation shows that $\varphi(\omega_{ij})\varphi(\omega_{jk}) + \varphi(\omega_{jk})\varphi(\omega_{ki}) + \varphi(\omega_{ki})\varphi(\omega_{ij}) = 0$. This map is clearly surjective in cohomology, as $H^*(Conf_{\mathbb{C}}(r))$ is generated in degree 1 and all the generators are in the image. Since the two CDGAs obviously have the same cohomology, we can thus deduce that φ is a quasi-isomorphism. This shows that $Conf_{\mathbb{C}}(r)$ is formal over $\mathbb{C}$, and thus over $\mathbb{Q}$ thanks to Theorem 2.96.   □

This proof does not work for $n \geq 3$. Indeed, it does not appear possible to find a direct quasi-isomorphism $H^*(Conf_{\mathbb{R}^n}(r)) \to \Omega_{dR}^*(Conf_{\mathbb{R}^n}(r))$ in general. Kontsevich showed the following theorem, by a more involved method (later refined by Lambrechts–Volić to take into account the real homotopy type):

Theorem 2.101 (Kontsevich [Kon99] and Lambrechts and Volić [LV14]) *The configuration spaces of* $\mathbb{R}^n$ *are formal for any* $n \geq 2$.

We will review the proof in more detail later (Sect. 5.4). Let us note already that the proof is much more complicated than for the case $n = 2$. It involves a zigzag:

$$H^*(Conf_{\mathbb{R}^n}(r)) \leftarrow \cdot \to \Omega_{PA}^*(Conf_{\mathbb{R}^n}(r)) \tag{2.64}$$

where the middle CDGA is an almost free resolution of the cohomology of $Conf_{\mathbb{R}^n}(r)$. Informally, the idea is to create a CDGA where there are no more relations between generators, but where the differential encodes the relations. To build the application into the CDGA of forms, we must no longer find forms that strictly satisfy the relations, but only forms that satisfy the relationship up to homotopy (i.e., the map is compatible with the differential).

It is this idea that we will use in the next chapter to deal with the case of configuration spaces of closed manifolds. We will also see that Kontsevich's result is much deeper. A certain algebraic (operadic) structure on configuration spaces is compatible with this formality. This compatibility has many applications, as we will see in Chap. 5.

Before moving on to the next chapter, let us already note that Theorem 2.101 allows us to glean a trove of information about the configuration spaces of $\mathbb{R}^n$. The explicit description of $H^*(\mathsf{Conf}_{\mathbb{R}^n}(r))$ already allowed us to compute the Betti numbers of $\mathsf{Conf}_{\mathbb{R}^n}(r)$, and we moreover know the cup product. Since the Massey products (Definition 2.30) of a formal space vanish (see [FOT08, Proposition 2.90]), we know that the Massey products of $\mathsf{Conf}_{\mathbb{R}^n}(r)$ vanish.

Furthermore, we can actually compute the rational homotopy groups of $\mathsf{Conf}_{\mathbb{R}^n}(r)$ from the explicit description of Theorem 2.88. It is more convenient to describe these homotopy groups as a Lie algebra under the Whitehead product (see Definitions 2.76 and 2.77).

Definition 2.102 Let $n \geq 2$ and $r \geq 0$ be integers. The *Drinfeld–Kohno Lie algebra* $\mathfrak{p}_n(r)$ is the free Lie algebra generated by symbols t_{ij} of degree $2 - n$, for $1 \leq i \neq j \leq n$, subject to the relations:

1. for distinct i and j, we have $t_{ji} = (-1)^n t_{ij}$;
2. for pairwise distinct i, j, k, we have $[t_{ik}, t_{ij} + t_{jk}] = 0$;
3. for pairwise distinct i, j, k, l, we have $[t_{ij}, t_{kl}] = 0$.

The following result can be seen through an explicit computation starting from Theorem 2.88, computing the minimal model, and using the isomorphism of Theorem 2.75:

Theorem 2.103 *There is an isomorphism of graded Lie algebras, for $n \geq 3$ and $r \geq 0$:*

$$\pi_{*+1}(\mathsf{Conf}_{\mathbb{R}^n}(r)) \otimes_{\mathbb{Z}} \mathbb{Q} \cong \mathfrak{p}_n(r). \tag{2.65}$$

Remark 2.104 Consistently with our convention stated at the beginning of the volume, we consider all degrees to be cohomological, so that $\mathfrak{p}_n(r)$ is non-positively graded (i.e., non-negatively with homological conventions). Homotopy groups being of dual nature to cohomology, they are also considered to be placed in negative cohomological degree.

Remark 2.105 In fact, more than that is true. For $n \geq 3$, the graded Lie algebra $\mathfrak{p}_n(r)$ actually forms a Quillen model of $\mathsf{Conf}_{\mathbb{R}^n}(r)$ (in the sense of the rational homotopy theory of Quillen [Qui69], see Remark 2.49). Since $\mathfrak{p}_n(r)$ has no differential, this can be interpreted as the fact that $\mathsf{Conf}_{\mathbb{R}^n}(r)$ is coformal, the dual notion of formality for Quillen models.

Remark 2.106 When $n = 2$ (i.e., we are looking at configurations in the plane), the elements $t_{ij} \in \pi_1(\mathsf{Conf}_{\mathbb{R}^2}(r))$ have a nice graphical interpretation. Recall that the fundamental group of $\mathsf{Conf}_{\mathbb{R}^2}(r)$ is the pure braid group PB_r (see Example 2.6), i.e., braids such that every strand starts and ends at the same position. Then t_{ij} can be represented by a braid such that the strand j crosses over the strand i exactly once, as in the following picture:

$$
t_{ij} =
$$

$$(2.66)$$

However, in that dimension, the isomorphism of Theorem 2.103 is not an isomorphism of Lie algebras but merely of graded vector spaces. In order to take the Lie algebra structure into account, one must take the Malcev completion of the fundamental group. In some sense, this means that $\mathfrak{p}_2(r)$ can be seen as an infinitesimal version of the pure braid group, see the work of Kohno [Koh85]. We refer to Fresse [Fre17a, Section 10] for a detailed treatment and a deep connection with the little disks operads and the Grothendieck–Teichmüller groups, and to Merkulov [Mer21] for a survey.

Chapter 3
Configuration Spaces of Closed Manifolds

In this chapter, we define the model conjectured by Lambrechts and Stanley [LS08a], and we show that their conjecture is true over $\mathbb{R}$ for a large class of closed manifolds.

The proof is an adaptation and a generalization of the formality of the configuration spaces of Euclidean spaces which was studied in Sect. 2.4. However, unlike Euclidean spaces, configuration spaces of closed manifolds are rarely formal. The conjectural model, denoted G_A and introduced in Sect. 3.1, must thus have a nontrivial differential. Moreover, it uses a model A of the manifold which has a special feature, namely, Poincaré duality.

After introducing the model, we introduce the compactifications of configuration spaces of Axelrod and Singer [AS94] and Fulton and MacPherson [FM94] in Sect. 3.2. These compactifications are needed because we are going to compute integrals on configuration spaces, and we need to ensure that these integrals converge. Moreover, the integral that we use are along the fibers of the projections between configuration spaces that forget some of the points in a configuration. Since these projections are not submersions any longer once we consider the compactifications [LV14, Example 5.9.1], we must use the theory of piecewise semi-algebraic forms [HLTV11; KS00] to properly define the integrals. This theory is introduced in Sect. 3.3.

Finally, we will explain how graph complexes, a combinatorial tool to provide resolutions of G_A, appear in the proof in Sect. 3.4. We also explain how a simple degree counting argument on graphs is essential in Sect. 3.5. As a corollary to this result, we obtain real homotopy invariance of configuration spaces of simply connected closed manifolds.

© The Author(s), under exclusive license to Springer Nature Switzerland AG 2022   47
N. Idrissi, *Real Homotopy of Configuration Spaces*, Lecture Notes
in Mathematics 2303, https://doi.org/10.1007/978-3-031-04428-1_3

3.1   The Lambrechts–Stanley Model

3.1.1   Definition of the Model

We start by introducing the Lambrechts–Stanley model of $\mathsf{Conf}_M(r)$. The idea behind this model is the following. By Poincaré–Lefschetz duality, if W is an oriented closed manifold of dimension n and $K \subset W$ is a compact subset which admits a neighborhood deformation retract whose complement is compact and locally contractible, then $H^*(W \setminus K) \cong H_{n-*}(W, K)$ [Hat02, Proposition 3.46]. The exact long sequence in homology tells us that $H_{n-*}(W, K)$ is obtained, so to speak, from the homology of W by "killing" the classes coming from K. To obtain the space $\mathsf{Conf}_M(r)$, we can start from the Cartesian product M^r and remove the diagonals $\Delta_{ij} = \{x \in M^r \mid x_i = x_j\}$. The model we are going to define is essentially built by applying Poincaré–Lefschetz duality to this description of $\mathsf{Conf}_r(M)$.

Let us start with some prerequisites. By Poincaré duality, if M is an oriented closed manifold, then there is a class $[M] \in H_n(M)$ such that for any $k \in \mathbb{Z}$, the pairing $H^k(M) \otimes H^{n-k}(M) \to \mathbb{R}$, $\alpha \otimes \beta \mapsto \langle \alpha\beta, [M] \rangle$ is non-degenerate. In particular, $H^k(M) = 0$ for $k > n$. We wish to generalize this at the "cochain level", rather than at the cohomology level. The following definition is a differential-graded version of commutative Frobenius algebras.

Definition 3.1   A *Poincaré duality CDGA* (of formal dimension n) is a pair (A, ε_A) where:

- A is a CDGA concentrated in nonnegative degrees and such that $A^0 = \mathbb{R}$;
- $\varepsilon_A : A^n \to \mathbb{R}$ is a linear map satisfying $\varepsilon(da) = 0$ for any $a \in A^{n-1}$ (which reflects the Stokes formula, as our manifold is without boundary);
- for any $k \in \mathbb{Z}$, the pairing

$$A^k \otimes A^{n-k} \to \mathbb{R}$$
$$a \otimes b \mapsto \varepsilon(ab)$$

$$\tag{3.1}$$

   is non-degenerate.

A *Poincaré duality model* of a closed manifold M is a model of M with a Poincaré duality CDGA structure.

Remark 3.2   Since we take $k \in \mathbb{Z}$ in the last statement, and since A is concentrated in nonnegative degrees, it follows that $A^k = 0$ for $k > n$. Moreover, ε_A must be an isomorphism between the top-degree component A^n and $\mathbb{R}$.

Remark 3.3   In the notation, one will often forget the ε_A.

Example 3.4   If a closed manifold M is formal, then $H^*(M)$ is a Poincaré duality model of it.

Example 3.5 There are Poincaré duality CDGAs which have nonzero differential. For example, consider $A = \bigl(S(x_2, y_3)/(x^3), dy = x^2\bigr)$ with $\varepsilon_A(x^2 y) = 1$. Note that this simple example is quasi-isomorphic to $H^*(A) = H^*(\mathbb{S}^2 \times \mathbb{S}^5)$. There even exist Poincaré duality CDGAs which are not quasi-isomorphic to any Poincaré duality CDGA with trivial differential. For example, we can apply the algorithm of Theorem 3.6 to [FOT08, Example 2.91]. We illustrate the first step of the algorithm in Example 3.8

Any model A of a manifold satisfies Poincaré duality at the level of cohomology. However, it does not necessarily satisfy it at the level of cochains. There is a naive idea to force Poincaré duality to hold at that level. Consider the ideal $O \subset A$ of elements $a \in A$ (called the orphans) for which the linear map $\varepsilon(a \cdot -) : A \to \mathbb{R}$ vanishes. In other words, we have:

$$O := \{a \in A \mid \forall b \in A, \ \varepsilon(ab) = 0\}. \tag{3.2}$$

Modding out by this ideal always yields a Poincaré duality CDGA A/O. This naive idea, however, does not work. The issue is that modding out by orphans can change the homotopy type of the CDGA, or said differently, the ideal O may fail to be acyclic. Indeed, an arbitrary element $a \in O$ is known to be a coboundary by Poincaré duality on cohomology; but it may be that it is not the coboundary of any orphan. However, Lambrechts and Stanley [LS08b] have managed to refine this idea into a procedure that works with any simply connected CDGA whose cohomology satisfies Poincaré duality.

Theorem 3.6 (Lambrechts and Stanley [LS08b]) *Let M be a simply connected closed manifold. There exists a model of M which is a Poincaré duality CDGA.*

***Proof** (**Idea of the Proof of Theorem 3.6**)* Let us now give an idea of the proof of Theorem 3.6 works. This will be helpful in Sect. 4.2, as we are going to explain how this proof generalizes to the case of manifolds with boundary. We are only going to explain the proof in characteristic zero. Note that positive characteristic presents an additional difficulty, pointed out below. In what follows, we can assume that $n \geq 7$. Otherwise, the space M is formal by Example 2.95.

Let A be a model of M, that we can assume to satisfy $A^0 = \mathbb{Q}$, $A^1 = 0$, $A^2 \subset \ker d$. Let us also fix some linear map $\varepsilon : A^n \to \mathbb{Q}$ such that $\varepsilon \circ d = 0$ and which induces the orientation on cohomology. Let us finally define the ideal $O \subset A$ as in Eq. (3.2). It is an easy exercise that A/O is a Poincaré duality CDGA. Our goal, in what follows, is to modify A so that the ideal of orphans in the modified CDGA is acyclic.

Following closely [LS08b], we say that O is k-half-acyclic if:

$$O^i \cap \ker d \subset d(O^{i-1}) \text{ for } n/2 + 1 \leq i \leq k, \tag{3.3}$$

i.e., the ideal O is acyclic between the degrees $n/2 + 1$ and k. The condition is vacuously satisfied when $k \leq n/2$. If the condition is satisfied for $k = n + 1$, then the upper half of O is acyclic, and so O is fully acyclic by Poincaré duality on cohomology. We are thus going to work our way up from $k = n/2$ and inductively reach a CDGA whose ideal of orphans is $(n + 1)$-half-acyclic.

Let us thus assume that the ideal of orphans is $(k - 1)$-half-acyclic. The obstruction to O being k-half-acyclic is given by $H^k(O)$. Let us thus choose a complement of $d(O^{k-1})$ in $O^k \cap \ker d$, and let us denote a basis of that complement by:

$$O^k \cap \ker d = d(O^{k-1}) \oplus \mathbb{Q}\langle \alpha_1, \dots, \alpha_l \rangle. \tag{3.4}$$

By Poincaré duality on cohomology, the α_i must be coboundaries, so let us choose elements γ_i' such that $d\gamma_i' = \alpha_i$. Note that the elements γ_i' cannot belong to the set of orphans, by definition.

Let us choose cocycles $\{h_i\}$ such that $\{[h_i]\}$ forms a basis of $H^*(A)$, and cocycles $\{h_i^\vee\}$ such that $\{[h_i^\vee]\}$ is the Poincaré dual basis of $\{h_i\}$ (i.e., $\varepsilon(h_i h_j^\vee) = \delta_{ij}$). Then we let:

$$\gamma_i := \gamma_i' - \sum_j \varepsilon(\gamma_j' h_j) h_j^\vee. \tag{3.5}$$

Then clearly $d\gamma_i = \alpha_i$, and $\varepsilon(\gamma_i \cdot a) = 0$ for any cocycle a (using Stokes' formula). We can thus define an extension of A by:

$$\hat{A} := \left(A \otimes S(c_1, \dots, c_l, w_1, \dots, w_l), d(c_i) = \alpha_i, d(w_i) = c_i - \gamma_i\right), \tag{3.6}$$

where $\deg c_i = k - 1$ and $\deg w_i = k - 2$. Note that since $n \geq 7$, we have $k \geq 5$. Then the canonical inclusion $A \cong A \otimes 1 \to \hat{A}$ is a quasi-isomorphism. Indeed, its cofiber is $\left(S(c_i, w_i), dc_i = 0, dw_i = c_i\right)$ which is obviously acyclic over $\mathbb{Q}$.

Remark 3.7 This is where the characteristic zero is essential. In characteristic $p > 0$, $d(w_i^p) = p w_i^{p-1} c_i = 0$ is a cocycle which is not the coboundary of anything. □

All that remains is to find an extension $\hat{\varepsilon} : \hat{A}^n \to \mathbb{Q}$ of $\varepsilon : A^n \to \mathbb{Q}$. The idea is to make it so that the c_i are orphans. Since their coboundaries are the α_i, this kills the obstruction to $\hat{A}$ being k-half-acyclic. And since k is high enough (thanks to $n \geq 7$), adding the new generators did not break the acyclicity in lower degrees. We refer to [LS08b] for the complete proof, or to the proof of Proposition 4.12 for a similar proof in the case where $\partial M \neq \emptyset$. □

Example 3.8 Let us illustrate how the Lambrechts–Stanley algorithm works on a simple example. Consider the space M of [FOT08, Example 2.91]. This space is a simply connected 7-manifold which sits in a fiber sequence $\mathbb{S}^3 \to M \to \mathbb{S}^2 \times \mathbb{S}^2$. It has a minimal model given by:

$$A = \left(S(a_2, b_2, u_3, v_3, t_3), da = db = 0, du = a^2, dv = b^2, dt = ab\right). \tag{3.7}$$

Its nonzero cohomology classes are spanned by the following classes:

- $H^0(M)$ is spanned by $[1]$;
- $H^2(M)$ is spanned by $[a]$ and $[b]$;
- $H^5(M)$ is spanned by $[ub - at]$ and $[av - tb]$;
- $H^7(M)$ is spanned by $[abt - a^2v] = [abt - b^2u]$;
- all other cohomology groups vanish.

Note that this manifold has two nonzero Massey products (see Definition 2.30), respectively given by:

$$\langle [a], [a], [b] \rangle = [ub - at], \qquad \langle [a], [b], [b] \rangle = [av - tb]. \qquad (3.8)$$

It follows that M cannot be formal, so the result of the algorithm will be a CDGA with a nonzero differential.

We choose as orientation $\varepsilon(abt) = -\varepsilon(a^2v) = -\varepsilon(b^2u) = 1/2$. We will illustrate the first two steps of the induction of Theorem 3.6. We begin at $k = \lceil n/2 + 1 \rceil = 5$. In degree 5, the CDGA A is spanned by au, at, bu, bt, av, and bv. Out of these, au and bv are orphans (as the pairings $\varepsilon(au \cdot x)$ and $\varepsilon(bv \cdot x)$ vanish for $x \in \{a, b\}$). However, since they are not cocycles, they do not constitute an obstruction to acyclicity

We may thus move onto the next step, i.e., $k = 6$. Consider the elements a^3, a^2b, ab^2, b^3, uv, ut, and vt. They are obviously all orphans since $A^1 = 0$, but the cocycles are just a^3, a^2b, ab^2, b^3. Only

$$\alpha_1 := a^2b \text{ and } \alpha_2 := ab^2 \qquad (3.9)$$

are obstructions to acyclicity of O, since $a^3 = d(au)$ and $b^3 = d(bv)$ are coboundaries of orphans. We have $a^2b = d(at)$ and $ab^2 = d(bt)$, where $\gamma_1' := at$ and $\gamma_2' := bt$ are not orphans (since $\varepsilon(at \cdot b) = \varepsilon(bt \cdot a) = 1/2 \neq 0$). To apply the algorithm, we choose $(h_j)_{j=1}^6 := (1, a, b, ub - at, av - tb, abt - a^2v)$ as cocycles which induce a basis of the cohomology. Moreover, we choose the cocycles which induce the dual basis $(h_j^\vee)_{j=1}^6 := (abt - a^2v, bt - av, at - bu, b, a, 1)$. We then get:

$$\gamma_1 = \gamma_1' - \sum_j \varepsilon(at \cdot h_j) h_j^\vee = at - \frac{1}{2}(at - bu) = \frac{1}{2}(at + bu), \quad \gamma_2 = \frac{1}{2}(bt + av).$$

$$(3.10)$$

Finally, we extend A to:

$$\hat{A} := \left(A \otimes S(c_1, c_2, w_1, w_2), d \right), \qquad (3.11)$$

$$d(c_1) = a^2b, \ d(w_1) = c_1 - \frac{1}{2}(at + bu), \qquad (3.12)$$

$$d(c_2) = ab^2, \ d(w_2) = c_2 - \frac{1}{2}(bt + av), \qquad (3.13)$$

where $\deg c_i = 5$ and $\deg w_i = 4$. It is an instructive exercise to try and determine what the extension $\hat{\varepsilon} : \bar{A}^7 \to \mathbb{Q}$ should be (see [LS08b, Equation (4.5)] for the answer). In order to finish the algorithm, we would need to continue the induction with $k = 7$ and finally $k = 8$, which we also leave as an exercise to the reader.

Armed with Theorem 3.6, we can assume that we are given a Poincaré duality model of our manifold M. Let us now quickly review some properties of Poincaré duality models.

Definition 3.9 Let A be a Poincaré duality CDGA. Let $\{a_i\}_{i \in I}$ be a graded basis of A, and let $\{a_i^\vee\}_{i \in I}$ be its dual basis (i.e. $\varepsilon(a_i a_j^\vee) = \delta_{ij}$ for any $i, j \in I$). The *diagonal class* of A is the element of $(A \otimes A)^n$ given by:

$$\Delta_A := \sum_{i \in I} (-1)^{|a_i|} a_i \otimes a_i^\vee. \tag{3.14}$$

This element is a cocycle of degree n and does not depend on the chosen basis. Using Sweedler's notation, we will often write $\Delta_A = \sum_{(\Delta_A)} \Delta_A' \otimes \Delta_A''$.

Remark 3.10 This class has a geometric interpretation. Let $[M] \in H_n(M)$ be the fundamental class of M. We push it forward along the diagonal map $\delta : x \mapsto (x, x)$ to get $\delta_*[M] \in H_n(M \times M)$. By Poincaré duality, this class corresponds to the diagonal class in $H^{2n-n}(M \times M)$.

Example 3.11 Let $A = H^*(\Sigma_g)$ be the cohomology of an oriented surface of genus g. It has a basis given by the elements $1 \in A^0$, $\alpha_1, \ldots, \alpha_g, \beta_1, \ldots, \beta_g \in A^1$ and $\upsilon = \alpha_i \beta_i \in A^2$. The augmentation is given by $\varepsilon(\upsilon) = 1$. The diagonal class of A is then:

$$\Delta_A = 1 \otimes \upsilon + \upsilon \otimes 1 - \sum_{i=1}^g (\alpha_i \otimes \beta_i + \beta_i \otimes \alpha_i). \tag{3.15}$$

Lemma 3.12 *The diagonal class satisfies* $(1 \otimes a)\Delta_A = (a \otimes 1)\Delta_A$ *for any* $a \in A$. *Furthermore, in Sweedler's notation, if* $\Delta_A = \sum_{(\Delta_A)} \Delta_A' \otimes \Delta_A''$, *then we have, for any* $x \in A$:

$$\sum_{(\Delta_A)} (-1)^{\deg \Delta_A'} \varepsilon(x \Delta_A'') \Delta_A' = x. \tag{3.16}$$

Remark 3.13 The first property is explained by the fact that one can represent Δ_A by a form on $M \times M$ whose support is concentrated along the diagonal $M \subset M \times M$ [MS74, Section 11, p. 125].

Proof Let a_i be a graded basis of A and $a_i^\vee$ be its dual basis as in Definition 3.9. Let $x \in A$ be an arbitrary element. Then $x = \sum_i \lambda_i a_i$ for some coefficients $\lambda_i \in \mathbb{R}$, and $\lambda_i = \varepsilon_A(x a_i^\vee)$. It follows that Eq. (3.16) holds. For the other formula, we have (with signs to write down carefully):

$$(1 \otimes a)\Delta_A = \sum_i \pm a_i \otimes aa_i^\vee$$

$$= \sum_i \pm a_i \otimes \left(\sum_j \pm \varepsilon_A(aa_i^\vee a_j)a_j^\vee \right)$$

$$= \sum_{i,j} \pm \varepsilon_A(aa_i^\vee a_j)a_i \otimes a_j^\vee$$

$$= \sum_j \pm aa_j \otimes a_j^\vee$$

$$= (a \otimes 1)\Delta_A.$$

$\square$

We will need the following notation.

Definition 3.14 Let A be a CDGA and $1 \leq i, j \leq r$ of integers. We define the morphism $p_i^* : A \to A^{\otimes r}$ by:

$$p_i^*(a) := 1 \otimes \cdots \otimes 1 \otimes \underbrace{a}_{\text{position } i} \otimes 1 \otimes \cdots \otimes 1. \tag{3.17}$$

We define moreover $p_{ij}^* : A \otimes A \to A^{\otimes r}$ by $p_{ij}^*(a \otimes b) = p_i^*(a) \cdot p_j^*(b)$.

Definition 3.15 (Lambrechts and Stanley [LS08a]) Let A be a Poincaré duality model of a simply connected closed manifold M. The *Lambrechts–Stanley model* associated to A is the CDGA:

$$\mathsf{G}_A(r) := \left(A^{\otimes r} \otimes S(\omega_{ij})_{1 \leq j \neq j \leq r} / I, d \right). \tag{3.18}$$

The dg-ideal I is generated by:

- the relations $\omega_{ij}^2 = 0$, $\omega_{ji} = (-1)^n \omega_{ij}$, and $\omega_{ij}\omega_{jk} + \omega_{jk}\omega_{ki} + \omega_{ki}\omega_{ij} = 0$ that appear in Theorem 2.88;
- and the symmetry relations $p_i^*(a)\omega_{ij} = p_j^*(a)\omega_{ij}$ for any $a \in A$ and $1 \leq i \neq j \leq r$.

The differential $d = d_A + d_{\text{split}}$ is the sum of the differential induced by that of A and the unique derivation that extends:

$$d_{\text{split}}(\omega_{ij}) = p_{ij}^*(\Delta_A). \tag{3.19}$$

Let us describe $\mathsf{G}_A(r)$ for small values of r. Recall the descriptions of $\mathsf{Conf}_M(r)$ for small values of r from Examples 2.2 and 2.3.

Example 3.16 The CDGA $G_A(0) = \mathbb{Q}$ is actually a model of $\mathsf{Conf}_M(0) = \{\varnothing\}$. Moreover, $G_A(1) = A$ is, by hypothesis, a model of $\mathsf{Conf}_M(1) = M$.

Example 3.17 The CDGA $G_A(2)$ is given by $\left(A \otimes S(\omega_{12}, \omega_{21})/I, d\right)$. The relation $\omega_{21} = (-1)^n \omega_{12}$ allows us to get rid of ω_{21}. Thanks to the relation $\omega_{12}^2 = 0$, we find that $G_A(2)$ splits as a direct sum:

$$G_A(2) \cong \left(\left((A \otimes A \otimes \mathbb{Q}1) \oplus (A \otimes A \otimes \mathbb{Q}\omega_{12})\right)/I, \, d\right). \tag{3.20}$$

The symmetry relation gives us $a \otimes 1 \otimes \omega_{12} = 1 \otimes a \otimes \omega_{12}$ for all $a \in A$. Since we consider the ideal generated by these relations, we thus have:

$$\begin{aligned} G_A(2) &\cong \left((A \otimes A \otimes \mathbb{Q}1) \oplus (A \otimes_A A \otimes \mathbb{Q}\omega_{12}), \, d\right) \\ &\cong \left((A \otimes A \otimes \mathbb{Q}1) \oplus (A \otimes \mathbb{Q}\omega_{12}), \, d\right). \end{aligned} \tag{3.21}$$

Finally, the differential is given by the sum of the differential of A and $d_{\text{split}}(a \otimes \omega_{12}) = (a \otimes 1)\Delta_A = (1 \otimes a)\Delta_A$. It thus follows that the CDGA $G_A(2) \cong \text{cone}(\delta)$ is the cone of the map:

$$\begin{aligned} \delta : A &\to A \otimes A \\ a &\mapsto (a \otimes 1)\Delta_A. \end{aligned} \tag{3.22}$$

Since δ is injective, the cone is quasi-isomorphic to the cokernel, i.e., to the quotient $(A \otimes A)/(\Delta_A)$. This echoes the classical result that $H^*(\mathsf{Conf}_M(2)) = H^*(M^2 \setminus \Delta) \cong H^*(M)^{\otimes 2}/(\Delta_M)$.

Starting from $r = 3$, the CDGA $G_A(r)$ no longer has such a simple description. Not all classes can be represented by elements coming from $A^{\otimes r}$. There may be non-trivial classes that need to involve the ω_{ij}.

We can graphically represent the elements of $G_A(r)$ just like we did for the elements of $H^*(\mathsf{Conf}_{\mathbb{R}^n}(r))$ after Theorem 2.88. A word $\omega_{i_1 j_1} \ldots \omega_{i_l j_l}$ corresponds to graphs with r numbered vertices, with edges $(i_1, j_1), \ldots, (i_l, j_l)$. Thanks to the Arnold relations, this graph is without double edges or loops. One can view the edges as unoriented/unordered but the sign is *a priori* undefined. The factor $A^{\otimes r}$ correspond to decorations of the r vertices. The decorations can move along edges thanks to the symmetry relation. Hence, one can view decorations as decorating connected components of the graph, rather than vertices.

The CDGA structure fits into this graphical description. The multiplication glues the graphs along their vertices and multiplies the decorations of the connected components thus merged. The differential is the sum of the differential of A (which acts as a derivation on the decorations of the connected components) and the sum of all the ways of performing the operation that we will informally call "cutting the edge". Cutting an edge consists in removing that edge from the graph (but keeping the incident vertices), and in multiplying the decorations of the incident connected components by the factors of $\Delta \in A \otimes A$.

Example 3.18 Consider the class of the graph with two vertices, one edge connecting the two, and the unique connected component decorated by $a \in A$. Its differential is the class of the graph with two vertices, no edges, and the two connected components decorated by the factors of

$$(a \otimes 1)\Delta_A = (1 \otimes a)\Delta_A \in A \otimes A.$$

Remark 3.19 This CDGA has already been extensively studied in one form or another. Let us quickly review some of its occurrences in the literature.

- Cohen and Taylor [CT78] have described a spectral sequence which converges to the cohomology of $\mathsf{Conf}_M(r)$ and whose page E^2 is precisely $\mathsf{G}_{H^*(M)}(r)$. This spectral sequence is, in fact, the Leray spectral sequence of the inclusion $\mathsf{Conf}_M(r) \subset M^r$. Interpreted in a modern way, the subsequent differentials (informally) encode the difference between $H^*(M)$ and a model A of M.
- Suppose that M is a smooth complex projective manifold. It is therefore a compact Kähler manifold, which is therefore formal according to Theorem 2.98. Kriz [Kri94] showed that in this case, $\mathsf{G}_{H^*(M)}(r)$ is indeed a rational model of $\mathsf{Conf}_M(r)$. His work is based on earlier results by Fulton and MacPherson [FM94] (themselves based on results of Morgan [Mor78] mentioned in Remark 2.99). At the same time, Totaro [Tot96] proved that, in this case, the Cohen–Taylor spectral sequence collapses after the page E^2. Thanks to a result of Deligne [Del75], this results in that $H^*(\mathsf{Conf}_r(M))$ is isomorphic to $H^*(\mathsf{G}_{H^*(M)}(r))$ as an algebra.
- Bendersky and Gitler [BG91] constructed a spectral sequence that converges to the relative cohomology $H^*(M^r, \Delta_M^{(r)})$, where

$$\Delta_M^{(r)} := \{x \in M^r \mid \exists i \neq j \text{ s.t. } x_i = x_j\} = M^r \setminus \mathsf{Conf}_M(r) \tag{3.23}$$

is the thick diagonal. By Poincaré–Lefschetz duality, this cohomology is isomorphic to the homology of $\mathsf{Conf}_M(r)$. Félix and Thomas [FT04], Berceanu et al. [BMP05], showed that the page E^2 of this spectral sequence is isomorphic to the dual of $\mathsf{G}_{H^*(M)}(r)$.
- Lambrechts and Stanley [LS04] have shown that if M is 2-connected, then $\mathsf{G}_A(2)$ is indeed a model of $\mathsf{Conf}_M(2)$. Cordova Bulens [Cor15] generalized this result to simply connected manifolds of even dimension.
- Lambrechts and Stanley [LS08a] have shown later on that if M is a simply connected closed manifold, then the cohomology of $\mathsf{G}_A(r)$ is isomorphic to the cohomology of $\mathsf{Conf}_M(r)$ as a representation of the symmetric group $\mathfrak{S}_r$, degree by degree.

3.1.2   Statement of the Theorem and Proof Strategy

The main goal of this chapter is to prove the following theorem:

Theorem 3.20 ([Idr19], See Also Campos and Willwacher [CW16]) *Let M be a simply connected smooth closed manifold, and let A be any Poincaré duality model of M. The Lambrechts–Stanley model $G_A(r)$ is a real model of $\mathrm{Conf}_M(r)$ for any $r \geq 0$.*

This thus settles Conjecture 2.86 for a large class of manifolds. To the authors' knowledge, little more is known about the non-smooth case.

The proof of Theorem 3.20 is an adaptation and generalization of Kontsevich's proof of the formality of configuration spaces of $\mathbb{R}^n$. The main steps are as follows.

1. It is (in general) impossible to find forms on $\mathrm{Conf}_M(r)$ that strictly satisfy the relations of $G_A(r)$. We thus start by constructing a resolution of $G_A(r)$ which is free as an algebra. This makes it possible to search only for forms that satisfy the relations up to homotopy. This resolution is constructed using decorated graph complexes as in Kontsevich's proof.
2. We show in a purely combinatorial way that this resolution is indeed quasi-isomorphic at $G_A(r)$. We proceed exactly as for the computation of the cohomology of $\mathrm{Conf}_{\mathbb{R}^n}(r)$.
3. To define the morphism $G_A(r) \to \Omega^*(\mathrm{Conf}_M(r))$, we want to use integrals, as in Kontsevich's proof. However, $\mathrm{Conf}_M(r)$ is not compact for $r \geq 2$, so integrals do not necessarily converge. We thus study the compactification of Axelrod–Singer–Fulton–MacPherson [AS94; FM94] of $\mathrm{Conf}_M(r)$.
4. The integrals in Kontsevich's proof are integrals along the fibers of the projections $\mathrm{Conf}_M(r + s) \to \mathrm{Conf}_M(r)$ that forget some points of the configuration. However, once the configuration spaces are compactified, these projections are no longer submersions. This prevents the use of the classical theory of de Rham differential forms. If M is semi-algebraic, these projections are however semi-algebraic fiber bundles, which allows us to use the theory of piecewise semi-algebraic forms [HLTV11; KS00]. Not all manifolds are semi-algebraic, but this is the case for smooth manifolds thanks to the theorem of Nash [Nas52] and Tognoli [Tog73].
5. A key point in the construction of graph complexes is reduction: it is necessary to quotient by certain graphs to have the right type of homotopy. However, it is generally not clear that the integration procedure along the fibers respects this quotient. An extremely simple counting argument shows that if $\dim M \geq 4$, then this is the case.

   Note that in dimension $\dim M \leq 3$, there are only three simply connected smooth closed manifolds: the point $\{0\}$, the sphere $\mathbb{S}^2$ (classification of surfaces) and $\mathbb{S}^3$ (Poincaré conjecture [KL08; MT07; Per02; Per03]). For each of these examples a different proof allows to show that the Lambrechts–Stanley model is a model of the configuration space.

6. It only remains to show that if A and B are two different Poincaré duality models of M, then $\mathsf{G}_A(r) \simeq \mathsf{G}_B(r)$. That proof involves graph complexes but remains purely algebraic (i.e., no integrals or assumptions about smoothness are needed).

Remark 3.21 In fact, we do not proceed in this order. The complete construction of the graph complex depends on the integrals, which themselves depend on the chosen compactification.

We saw in Remark 2.91 that for simply connected manifolds of dimension at least 3, then $\mathsf{Conf}_M(r)$ is simply connected. It follows that real models of $\mathsf{Conf}_M(r)$ (in the sense that we have defined here) encode its real homotopy type. The case of dimension ≤ 2 is trivial in that regard, as homotopy equivalent manifolds are actually homeomorphic in these dimensions. Therefore, using Theorem 3.20, we obtain the following corollary:

Corollary 3.22 *Let M and N be two manifolds satisfying the hypotheses of the theorem. If they have the same real homotopy type, then so do their configuration spaces.*

It is not a priori obvious that if A and B are two Poincaré duality CDGAs that are quasi-isomorphic, then $\mathsf{G}_A(r)$ and $\mathsf{G}_B(r)$ are quasi-isomorphic. This is the case if we can find a direct quasi-isomorphism of CDGA $f : A \to B$ such that $\varepsilon_B \circ f = \varepsilon_A$. It should be noted, however, that the existence of such a quasi-isomorphism is restrictive, as can be seen by the following lemma.

Lemma 3.23 *Let A and B be two connected Poincaré duality CDGAs. If $f : A \to B$ is a quasi-isomorphism, then f is injective.*

Proof Let us first note that A and B must have the same formal dimension n, as it is the largest degree for which the cohomology is nonzero. It is moreover easy to see that $H^n(A) = A^n = \mathbb{R}$ and $H^n(B) = B^n = \mathbb{R}$ (with ε playing the role of an isomorphism), so $f : A^n \to B^n$ is multiplication by a nonzero scalar λ. Suppose that $a \in A^k$ is a nonzero element. Then by Poincaré duality, there exists some $a' \in A^{n-k}$ such that $\varepsilon_A(aa') = 1$. It follows that

$$\varepsilon_B(f(a)f(a')) = \varepsilon_B(f(aa')) = \lambda \neq 0, \tag{3.24}$$

so $f(a) \neq 0$.   □

Lambrechts and Stanley [LS08b] have shown that if $n \geq 7$, $A^0 = B^0 = \mathbb{R}$, $A^1 = A^2 = B^1 = B^2 = 0$ and $H^3(A) = H^3(B) = 0$, then there is a zigzag of quasi-isomorphisms of Poincaré duality CDGAs $A \to C \leftarrow B$. This was recently improved by Hájek [Háj20] to simply connected CDGAs with vanishing H^2 of any formal dimension. But these conditions are still restrictive.

3.2 Fulton–MacPherson Compactifications

Unless $\dim M = 0$, the configuration spaces $\mathsf{Conf}_M(r)$ are not compact for $r \geq 2$, even if M is. The proof of Theorem 3.20 involves integrals on configuration spaces. To ensure that these integrals converge, one possibility is to compactify the configuration spaces.

In this section, we define a manifold with corners $\mathsf{FM}_M(r)$ whose interior is $\mathsf{Conf}_M(r)$. This manifold was initially defined by Fulton and MacPherson [FM94] in the complex setting (using blowups with projective spaces instead of spheres) and Axelrod and Singer [AS94] in the real setting (using spherical blowups). It was then studied in detail in [Sin04]. The boundary points of $\mathsf{FM}_M(r)$ informally consist of "virtual" configurations of points, where some points are infinitesimally close to each other. To obtain a correct homotopy type, one keeps local information about these point clusters in these virtual configurations. This local information essentially consists of a configuration (which can itself be virtual) in the tangent space of M.

In what follows, for ease of reasoning, it will be convenient to index the points of a configuration by elements of any finite set, rather than $\{1, \dots, r\}$.

Definition 3.24 Let U be a finite set. We define $\mathsf{Conf}_M(U)$ as the set of injections $U \hookrightarrow M$, seen as a subset of M^U.

Remark 3.25 The configuration space $\mathsf{Conf}_M(r)$ is simply $\mathsf{Conf}_M(\{1, \dots, r\})$. We will switch between the two notation indiscriminately.

3.2.1 Case of Euclidean Spaces

An important ingredient of the proof of Theorem 3.20 will be the Stokes formula, which will allow us to verify that our integration procedure preserves the differential. We will therefore describe the boundary of $\mathsf{FM}_M(r)$, and more generally the fiberwise boundary of the canonical projection $\mathsf{FM}_M(r+s) \to \mathsf{FM}_M(r)$. This description involves the compactifications of $\mathsf{Conf}_{\mathbb{R}^n}(r)$, so this is where we will start. The results of this section come from [LV14, Chapter 5] unless otherwise indicated.

Definition 3.26 Let U be a finite set and $i \neq j \neq k \neq i \in U$ three pairwise distinct elements. We define maps:

$$\theta_{ij} : \mathsf{Conf}_{\mathbb{R}^n}(U) \to \mathbb{S}^{n-1}, \qquad \delta_{ijk} : \mathsf{Conf}_{\mathbb{R}^n}(U) \to [0, +\infty],$$

$$x \mapsto \frac{x_i - x_j}{\|x_i - x_j\|}, \qquad\qquad x \mapsto \frac{\|x_i - x_k\|}{\|x_j - x_k\|}. \qquad (3.25)$$

The group of translations and positive dilations $\mathbb{R}^n \rtimes \mathbb{R}_{>0}$ acts on $\mathsf{Conf}_{\mathbb{R}^n}(U)$. If $\#U \geq 2$, this action is free and proper. The quotient $\mathsf{Conf}_{\mathbb{R}^n}(U)/\mathbb{R}^n \rtimes \mathbb{R}_{>0}$ is thus still a manifold, of dimension $n\#U - n - 1$. We can represent the elements of the quotient (for $\#U \geq 2$) by normalized configurations, i.e., those configurations $x \in \mathsf{Conf}_{\mathbb{R}^n}(r)$ whose barycenter $\sum_i \frac{1}{r} x_i$ is at the origin $0 \in \mathbb{R}^n$ and whose radius $\max(\|x_i\|)_{1 \leq i \leq r}$ is 1. If $\#U \leq 1$, the action is transitive and the quotient is thus reduced to a point. In both cases, the projection $\mathsf{Conf}_{\mathbb{R}^n}(U) \to \mathsf{Conf}_{\mathbb{R}^n}(U)/\mathbb{R}^n \rtimes \mathbb{R}_{>0}$ is a homotopy equivalence.

The maps of Definition 3.26 are compatible with the quotient by $\mathbb{R}^n \rtimes \mathbb{R}_{>0}$. They define an embedding, where $\mathsf{Conf}_U(2) = \{(i, j) \in U^2 \mid i \neq j\}$ and $\mathsf{Conf}_U(3) = \{(i, j, k) \in U^3 \mid i \neq j \neq k \neq i\}$:

$$\mathsf{Conf}_{\mathbb{R}^n}(U) \,/\, \mathbb{R}^n \rtimes \mathbb{R}_{>0} \hookrightarrow \left(\mathbb{S}^{n-1}\right)^{\mathsf{Conf}_U(2)} \times [0, +\infty]^{\mathsf{Conf}_U(3)}. \tag{3.26}$$

Definition 3.27 The *(Axelrod–Singer–)Fulton–MacPherson compactification* of $\mathsf{Conf}_{\mathbb{R}^n}(U)$, denoted $\mathsf{FM}_n(U)$, is the closure of the image of the embedding (3.26).

Recall that a manifold with corners is a smooth manifold which is locally diffeomorphic to open subsets of $[0, 1]^n$.

Proposition 3.28 ([GJ94], [LV14, Proposition 5.2]) *The space $\mathsf{FM}_n(U)$ is a compact manifold with corners. Its interior is the space $\mathsf{Conf}_{\mathbb{R}^n}(U)/\mathbb{R}^n \rtimes \mathbb{R}_{>0}$. Its dimension is $n\#U - n - 1$ for $\#U \geq 2$, and it is zero-dimensional otherwise.*

Proof Explicit charts have been described by Sinha [Sin04], where $\mathsf{FM}_n(U)$ is denoted $\tilde{C}_{\#U}(\mathbb{R}^n)$ and the charts are the $v_T^{\mathbb{R}^n, x}$ of [Sin04, Definition 3.33], projected down to $\mathsf{FM}_n(U)$. □

These compactifications have beautiful geometrical descriptions.

Example 3.29 The spaces $\mathsf{FM}_n(0)$ and $\mathsf{FM}_n(1)$ are singletons. The space $\mathsf{FM}_n(2)$ is just the unit sphere $\mathbb{S}^{n-1}$: up to translations, one of the two points can be fixed at the origin, and up to rescaling, the radius of the configuration can be fixed to 1.

Example 3.30 ([Hoe12, Figure 2]) The space $\mathsf{FM}_2(3)$ is a 3-manifold with boundary. It is obtained by removing a tubular neighborhood of three interlinked circles from the 3-sphere $\mathbb{S}^3$. Alternatively, it can be describe as a solid torus from which a tubular neighborhood of a Hopf linked (two interlinked circles) has been removed. See Fig. 3.1.

Example 3.31 Let $r \geq 0$. The space $\mathsf{FM}_1(r)$ has $r!$ connected components, each indexed by a permutation of $\{1, \ldots, r\}$, which corresponds to the various ways of ordering a configuration of r points on the real lines. The symmetric group (which acts by renumbering) simply permutes the connected components, so we may study only one of them, say the component of the identity permutation $K_r \subset \mathsf{FM}_1(r)$.

The space K_r is a compact manifold with corners of dimension $r - 2$ (for $r \geq 2$) called the rth associahedron. Associahedra were introduced by Stasheff [Sta61] after earlier work of Tamari [Tam51]. The space K_r can be realized as a convex polyhedron (and are also called Stasheff polytopes). The vertices of K_r

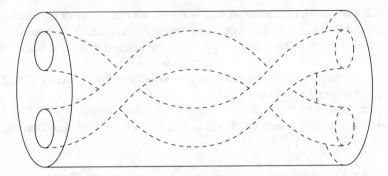

Fig. 3.1 The space $FM_2(3)$ is obtained from a solid torus (i.e., the cylinder with two opposite ends identified) by removing two interlinked solid tori (Figure adapted from [Hoe12, Figure 2])

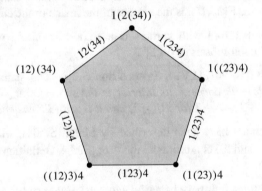

Fig. 3.2 The space $FM_1(4)$ is a disjoint union of 24 pentagons such as this one

correspond to possible parenthesizations of the word $x_1 \ldots x_r$ (where the x_i are arbitrary variables). Two parenthesizations are connected by an edge if one can be obtained from the other by applying the associativity rule exactly once. For example, the space $FM_1(3)$ is a union of six segments: the two vertices correspond to the two parenthesizations of $x_i x_j x_k$ (where (i, j, k) is some permutation of $(1, 2, 3)$), and the edge to the associativity rule. The space $FM_1(4)$ is a union of 24 pentagon (see Fig. 3.2): the vertices correspond to the four possible parenthesization of the word $x_i x_j x_k x_l$ (for some permutation (i, j, k, l) of $(1, 2, 3, 4)$), the edges to the possible applications of the associativity rule, and the interior corresponds to a coherence relation between the various applications of the associativity. Using the theory of operads (see Chap. 5), this intuitive idea can be made precise.

Like all manifolds with corners, $FM_n(U)$ deformation retracts on its interior. It follows that:

Corollary 3.32 ([Sal01, Proposition 2.5]) *The map* $\mathsf{Conf}_{\mathbb{R}^n}(U) \to FM_n(U)$, *given by the composition of the quotient by the* $\mathbb{R}^n \rtimes \mathbb{R}_{>0}$ *action and inclusion, is a homotopy equivalence.*

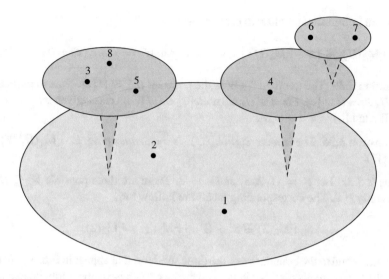

Fig. 3.3 An element of $FM_2(8)$

One can also characterize the boundary of that manifold.

Proposition 3.33 *The boundary of* $FM_n(U)$ *is characterized by elements where at least one of the coordinates* δ_{ijk} *vanishes.*

If a configuration $x \in FM_n(U)$ satisfies $\delta_{ijk}(x) = 0$, we will say that x_i is *infinitesimally close* to x_j with respect to x_k.

One can imagine that a point of $FM_n(U)$ is a normalized configuration of $\mathbb{R}^n$ which is potentially "virtual": some points can be infinitesimally close to each other. Such proximity relations can be nested. Figure 3.3 gives an example, where infinitesimal proximity relations are represented by smaller "zoomed in" disks:

- points 3, 5 and 8 are infinitesimally close to each other in relation to the others (i.e., we have $\delta_{358}(x) = 0$ for $x \notin \{3, 5, 8\}$, etc.);
- points 4, 6, and 7 are also infinitesimally close to each other;
- Moreover, points 6 and 7 are infinitesimally close to each other with respect to point 4.

The manifold $FM_n(U)$ is a manifold with corners, so its boundary $\partial FM_n(U)$ is itself a manifold with corners, the boundary of its boundary can itself have a boundary, etc. Let us now describe the facets of the boundary, i.e., the decomposition of the boundary into submanifolds of codimension zero which intersect along their own boundaries.

Definition 3.34 Let U be a finite set and let $W \subsetneq U$ be a subset of cardinality $\#W \geq 2$. We define the subset of virtual configurations such that all points indexed by W are infinitesimally close to each other:

$$\partial_W \mathsf{FM}_n(U) := \big\{ x \in \mathsf{FM}_n(U) \mid i, k \in W \wedge j \notin W \implies \delta_{ijk}(x) = 0 \big\}. \qquad (3.27)$$

Definition 3.35 The quotient U/W is the finite set $U \setminus W \sqcup \{*\}$. Note in particular that $U/\varnothing = U \sqcup \{*\}$. For $u \in U$, we note $[u] \in U/W$ its class, given by $[u] = u$ if $u \notin W$, and $[u] = *$ if $u \in W$.

Proposition 3.36 *The space* $\partial_W \mathsf{FM}_n(U)$ *is homeomorphic to* $\mathsf{FM}_n(U/W) \times \mathsf{FM}_n(W)$.

Example 3.37 Let $U = \{1, 2, 3\}$ and $n = 2$. There are three possible $W \subsetneq U$ of cardinality ≥ 2. The corresponding facets are hollow tori,

$$\partial_W \mathsf{FM}_2(3) \cong \mathbb{S}^1 \times \mathbb{S}^1 \cong \mathsf{FM}_2(2) \times \mathsf{FM}_2(2), \qquad (3.28)$$

that correspond to the boundaries of the three solid tori that appear in Fig. 3.1. In this case, the facets are disjoint, but in general the facets can have nonempty intersection as the next example shows.

Example 3.38 Let $U = \{1, 2, 3, 4\}$ and $n = 1$. Recall from Example 3.31 that $\mathsf{FM}_1(r) \cong \mathfrak{S}_r \times K_r$ where $\mathfrak{S}_r$ is the symmetric group and K_r is a connected polyhedron. In particular, K_2 is a singleton, $K_3 = \mathbb{D}^1$ is an interval, and K_4 is a pentagon. There are ten possible $W \subsetneq U$ of cardinality ≥ 2. The facets are either homeomorphic to $\mathsf{FM}_1(2) \times \mathsf{FM}_1(3) \cong \mathfrak{S}_2 \times \mathfrak{S}_3 \times \mathbb{D}^1$ if $\#W = 3$, or to $\mathsf{FM}_1(3) \times \mathsf{FM}_1(2) \cong \mathfrak{S}_3 \times \mathbb{D}^1 \times \mathfrak{S}_2$ if $\#W = 2$. (Although the two possible kinds of facets are homeomorphic, we view them as distinct. Precise indexing of points make the matter clearer.) Let us consider their intersection with the connected component where the four points are in the natural order (i.e., $x_1 < x_2 < x_3 < x_4$). Then we find the following intersections, as represented in Fig. 3.2:

- The facet $\partial_{\{1,2\}}\mathsf{FM}_2(4)$ intersects as the edge decorated by $(12)34$;
- The facet $\partial_{\{2,3\}}\mathsf{FM}_2(4)$ intersects as the edge decorated by $1(23)4$;
- The facet $\partial_{\{3,4\}}\mathsf{FM}_2(4)$ intersects as the edge decorated by $12(34)$;
- The facet $\partial_{\{1,2,3\}}\mathsf{FM}_2(4)$ intersects as the edge decorated by $(123)4$;
- The facet $\partial_{\{2,3,4\}}\mathsf{FM}_2(4)$ intersects as the edge decorated by $1(234)$;
- The other facets have empty intersection with that connected component.

Note on this example that the facets, which are of codimension 1, intersect along submanifolds of codimension 2 if they intersect at all.

Proof (Proposition 3.36) Let us build a map:

$$\circ_W : \mathsf{FM}_n(U/W) \times \mathsf{FM}_n(W) \to \mathsf{FM}_n(U). \qquad (3.29)$$

Let $x = (x_{[u]})_{[u] \in U/W} \in \mathsf{FM}_n(U/W)$ and $y = (y_w)_{w \in W} \in \mathsf{FM}_n(W)$ be two configurations. We define $x \circ_W y \in \mathsf{FM}_n(U)$ in the coordinate system $(\theta_{ij}, \delta_{ijk})$ by:

$$\theta_{ij}(x \circ_W y) = \begin{cases} \theta_{ij}(y), & \text{if } i, j \in W; \\ \theta_{[i][j]}(x), & \text{otherwise.} \end{cases} \tag{3.30}$$

$$\delta_{ijk}(x \circ_W y) = \begin{cases} \delta_{ijk}(y), & \text{if } i, j, k \in W; \\ 1, & \text{if } i, j \in W \text{ and } k \notin W; \\ +\infty, & \text{if } i \notin W \text{ and } j, k \in W; \\ 0, & \text{if } i, k \in W \text{ and } j \notin W; \\ \delta_{[i][j][k]}(x), & \text{otherwise.} \end{cases} \tag{3.31}$$

It is easily verified that the image of $\circ_W$ is equal to $\partial_W \mathsf{FM}_n(U)$ and that the application is injective. The result can thus be established by appealing to compactness of $\mathsf{FM}_n(U/W) \times \mathsf{FM}_n(W)$. □

Remark 3.39 The maps appearing in this proof are part of the operadic structure of FM_n, which we will study in Chap. 5 (see Proposition 5.59).

Proposition 3.40 *The boundary of* $\mathsf{FM}_n(U)$ *is the union of the facets* $\partial_W \mathsf{FM}_n(U)$. *More precisely, if we set* $\mathcal{BF}(U) = \{W \subsetneq U \mid \#W \geq 2\}$, *we then have:*

$$\partial \mathsf{FM}_n(U) = \bigcup_{W \in \mathcal{BF}(U)} \partial_W \mathsf{FM}_n(U). \tag{3.32}$$

Moreover, $\operatorname{codim} \partial_W \mathsf{FM}_n(U) = 1$, *and, if* $W \neq W'$, *then* $\operatorname{codim} \partial_W \mathsf{FM}_n(U) \cap \partial_{W'} \mathsf{FM}_n(U) > 1$.

Proof The fact that the boundary is equal to the union is immediate thanks to the fact that $x \in \partial \mathsf{FM}_n(U) \iff \exists i \neq j \neq k \neq i$ such that $\delta_{ijk}(x) = 0$. The fact that $\operatorname{codim} \partial_W \mathsf{FM}_n(U) = 1$ comes from the fact that $\circ$ is a homeomorphism and from the computation of the dimension of $\mathsf{FM}_n(U/W) \times \mathsf{FM}_n(W)$. Finally, the computation of the codimension of the intersection is a small exercise (there are several cases to check: $W \cap W' = \varnothing$, $W \subset W'$ or the reverse, or $W \cap W' \neq \varnothing$ but no inclusion). □

Recall that our goal is to apply the Stokes formula. To do this, we need to know the fiberwise boundary of the canonical projections.

Definition 3.41 Let $\pi : E \to B$ an oriented fiber bundle (i.e. the fibers are compact manifolds with compatible orientations). Its *fiberwise boundary* is the fiber bundle $\pi^\partial : E^\partial \to B$ where E^∂ is defined by:

$$E^\partial := \bigcup_{b \in B} \partial \pi^{-1}(b). \tag{3.33}$$

Example 3.42 Let us consider the fiber bundle $\pi : [0, 1]^2 \to [0, 1]$ which projects on the first coordinate. Its fiberwise boundary is $[0, 1] \times \{0, 1\}$. We notice that it is not the boundary of the total space or the pre-image of the boundary of the base.

Proposition 3.43 *Let* $U \subset A$ *be a pair of finite sets. The projection* $\pi :$ $\mathsf{Conf}_{\mathbb{R}^n}(A) \to \mathsf{Conf}_{\mathbb{R}^n}(U)$ *which forgets some points extends to an oriented fiber bundle:*

$$\pi : \mathsf{FM}_n(A) \to \mathsf{FM}_n(U). \tag{3.34}$$

Proof Let $x \in \mathsf{Conf}_{\mathbb{R}^n}(A)$ be some possibly virtual configuration. We simply need to define $\pi(x)$ in the coordinates of Definition 3.26. We can just take, for $i \neq j \neq k \neq i \in U$,

$$\theta_{ij}(\pi(x)) := \theta_{ij}(x), \quad \delta_{ijk}(\pi(x)) := \delta_{ijk}(x). \tag{3.35}$$

□

Proposition 3.44 *Let* $U \subset A$ *be a pair of finite sets. Let* $\mathcal{BF}(A, U) = \{W \in \mathcal{BF}(A) \mid U \subset A$ *or* $\#(W \cap A) \leq 1$. *Then the fiberwise boundary* $\pi : \mathsf{FM}_n(A) \to \mathsf{FM}_n(U)$ *is given by:*

$$\mathsf{FM}_n^\partial(A) = \bigcup_{W \in \mathcal{BF}(A,U)} \partial_W \mathsf{FM}_n(A). \tag{3.36}$$

Proof As π is a fiber bundle, it is simply a matter of checking which facets $\partial_W \mathsf{FM}_n(A)$ are sent to the interior of $\mathsf{FM}_n(U)$ via the projection. Indeed, $\mathsf{FM}_n^\partial(A)$ is the closure of $\partial \mathsf{FM}_n(A) \cap \pi^{-1}(\mathring{\mathsf{FM}}_n(U))$. One easily verifies that these are indeed the facets $W \in \mathcal{BF}(A, U)$. □

3.2.2　Case of Closed Manifolds

We can now do the same work for a closed manifold M. Thanks to Whitney's theorem, we can embed M in some Euclidean space $\mathbb{R}^N$ for N big enough. In what follows, we fix such an embedding and we see implicitly any element of M as a vector of $\mathbb{R}^N$. Many of the proofs will be omitted, as they are almost identical to the case of Euclidean spaces.

Definition 3.45 Let U a finite set and $i, j, k \in U$ three pairwise distinct elements. We define maps (by abuse of notation, we keep the same letters as in the previous subsection):

$$\theta_{ij} : \mathsf{Conf}_M(U) \to \mathbb{S}^{N-1}, \qquad \delta_{ijk} : \mathsf{Conf}_M(U) \to [0, +\infty],$$

$$(x_u)_{u \in U} \mapsto \frac{x_i - x_j}{\|x_i - x_j\|}, \qquad (x_u)_{u \in U} \mapsto \frac{\|x_i - x_k\|}{\|x_j - x_k\|}. \tag{3.37}$$

These map, together with the obvious inclusion $(p_u)_{u \in U} : \mathsf{Conf}_M(U) \hookrightarrow M^U$, define an embedding:

$$\mathsf{Conf}_M(U) \hookrightarrow M^U \times (\mathbb{S}^{N-1})^{\mathsf{Conf}_U(2)} \times [0, +\infty]^{\mathsf{Conf}_U(3)}. \tag{3.38}$$

Definition 3.46 The Fulton–MacPherson compactification of $\mathsf{Conf}_M(U)$, denoted by $\mathsf{FM}_M(U)$, is the closure of the image of the embedding (3.38).

We have the following theorem, similar to Proposition 3.28:

Proposition 3.47 *Let M be a compact smooth n-manifold and U be a finite set. The space $\mathsf{FM}_M(U)$ is a compact smooth manifold with corners of dimension $n\#U$.*

We define as previously the facets $\partial_W \mathsf{FM}_M(U)$, for $W \subset U$, as the set of configurations where the points indexed by W are infinitesimally close to each other compared to the points not in W:

$$\partial_W \mathsf{FM}_M(U) = \left\{ x \in \mathsf{FM}_M(U) \,\middle|\, i, k \in W \implies \begin{matrix} p_i(x) = p_k(x), \text{ and} \\ j \notin W \implies \delta_{ijk}(x) = 0 \end{matrix} \right\}. \tag{3.39}$$

Proposition 3.48 *The facet $\partial_W \mathsf{FM}_M(U)$ is a fiber bundle over $\mathsf{FM}_M(U/W)$, with fiber $\mathsf{FM}_n(W)$.*

Example 3.49 Let $U = W = \{1, 2\}$. Then $\partial_W \mathsf{FM}_M(2) = \partial \mathsf{FM}_M(2)$ is a sphere bundle over $\mathsf{FM}_M(1)$. This bundle is, in fact, the sphere bundle associated to the tangent bundle of M.

Proposition 3.50 *The boundary of $\mathsf{FM}_M(U)$ is given by:*

$$\partial \mathsf{FM}_M(U) = \bigcup_{W \in \mathcal{BF}_M(U)} \partial_W \mathsf{FM}_M(U), \tag{3.40}$$

where $\mathcal{BF}_M(U) = \{W \subset U \mid \#W \geq 2\}$. These facets are of codimension 1 and the intersection of two different facets is of codimension > 1.

Remark 3.51 Contrary to the case of $\mathbb{R}^n$, the facet $\partial_U \mathsf{FM}_M(U)$ is included in the boundary. This corresponds to the case where all points are at the same location in M. This cannot happen in $\mathbb{R}^n$ because of the quotient by positive dilations: if all points converge at the same speed to the same position, then up to dilation they do not actually move.

Proposition 3.52 *Let $U \subset A$ a pair of finite sets. The projection $\pi : \mathsf{Conf}_M(A) \to \mathsf{Conf}_M(U)$ which forgets some points extends to the compactification.*

Proposition 3.53 *The fiberwise boundary of $\pi : \mathsf{FM}_M(A) \to \mathsf{FM}_M(U)$ is given by:*

$$\mathsf{FM}_M^\partial(A) = \bigcup_{W \in \mathcal{BF}_M(A,U)} \partial_W \mathsf{FM}_M(A), \tag{3.41}$$

where $\mathcal{BF}_M(A, U) = \{W \in \mathcal{BF}_M(A) \mid \#(W \cap U) \leq 1\}$.

3.3   Semi-algebraic Sets and PA Forms

The canonical projections $\pi : \mathsf{FM}_M(A) \to \mathsf{FM}_M(U)$ are unfortunately not submersions. An example is given by Lambrechts and Volić [LV14, Example 5.9.1], which we now quickly describe. Recall from Example 3.31 that the space $\mathsf{FM}_1(3)$ is given by $3! = 6$ segments: the extremities of each segment are the two ways of parenthesizing a permutation of the word 123, and the path between the extremities is an associator. Moreover, the space $\mathsf{FM}_1(4)$ is given by $4! = 24$ pentagons. The vertices of each pentagon are the five ways to parenthesize a permutation of the word 1234, and the edges are paths that involve an associator. The projection $\mathsf{FM}_1(4) \to \mathsf{FM}_1(3)$ resembles, on the connected components in question, Fig. 3.4. A calculation in a smooth chart shows that π is not a submersion at the point corresponding to $1((23)4)$.

As these maps are not submersions, it is not possible to apply the standard theory of integration along the fibers of differential forms. However, these projections happen to be *semi-algebraic* fiber bundles. Initially developed by Kontsevich and Soibelman [KS00], the theory of piecewise semi-algebraic forms has been refined

Fig. 3.4   The projection $\mathsf{FM}_1(4) \to \mathsf{FM}_1(3)$ is not a submersion at the vertex labeled by $1((23)4)$

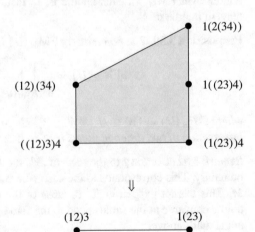

by Hardt et al. [HLTV11] in order to apply it to the proof of the formality of the operad of small discs. Let us now quickly recall the main ingredients of this theory. Unless otherwise indicated, we refer to [HLTV11] for the definitions and results of this section.

3.3.1 Semi-algebraic Sets

Definition 3.54 A *semi-algebraic (SA) set* is a subset of $\mathbb{R}^N$ (for a certain N) which is a finite union of finite intersections of sets of solutions of polynomial inequalities. An *SA map* is a (continuous) map between two SA sets whose graph is a semi-algebraic set. An *SA manifold* (with corners) of dimension n is an SA set locally homeomorphic to $\mathbb{R}^n$ or $\mathbb{R}_+ \times \mathbb{R}^{n-1}$.

Proposition 3.55 *The Fulton–MacPherson compactifications* $\mathsf{FM}_n(k)$ *and* $\mathsf{FM}_M(k)$ *are semi-algebraic manifolds for any* $n, k \geq 0$ *and any semi-algebraic manifold* M.

Proof The embeddings $(\theta_{ij}, \delta_{ijk})$ and $(\iota, \theta_{ij}, \delta_{ijk})$ are clearly SA maps, so their images and thus the closure of their images are SA sets. $\qquad\square$

Definition 3.56 An *SA fiber bundle* is an SA map $\pi : E \to B$ together with an SA set F, a cover $B = \bigcup\{U_\alpha\}$ by SA subsets, and SA homeomorphisms $h_\alpha : U_\alpha \times F \cong \pi^{-1}(U_\alpha)$ compatible with π.

Theorem 3.57 (Lambrechts and Volić [LV14]) *The canonical projection* $\pi :$ $\mathsf{FM}_n(A) \to \mathsf{FM}_n(U)$ *is an SA fiber bundle for any pair of finite sets* $U \subset A$.

The proof is relatively difficult. The fibers are represented by discs from which balls have been removed. The homeomorphisms must then be constructed explicitly, which is quite technical. The proof of the following result is almost identical to that of Lambrechts and Volić [LV14]. It is in fact slightly simpler, because there is no quotient by the group $\mathbb{R}^n \rtimes \mathbb{R}_{>0}$.

Proposition 3.58 *The canonical projection* $\pi : \mathsf{FM}_M(A) \to \mathsf{FM}_M(U)$ *is an SA fiber bundle for any pair of finite sets* $U \subset A$.

3.3.2 Piecewise Semi-algebraic Forms

We now introduce the notion of piecewise semi-algebraic forms. The definition is quite involved and goes through differential geometry and two duality steps (from differential forms to SA chains to a subset of SA cochains). We will skip the most technical parts of [HLTV11]. The reader might moreover want to skip this section at first, simply recording Theorem 3.74, Proposition 3.78 and Proposition 3.79.

Definition 3.59 A *current* of degree k on $\mathbb{R}^N$ is an element of the dual of the differential forms of degree k on $\mathbb{R}^n$ with compact support. The space of all currents is denoted by $D_k(\mathbb{R}^N) := \Omega_c^k(\mathbb{R}^N)^\vee$. A *current* of degree k on an SA set $X \subset \mathbb{R}^N$ is a current $T \in D_k(\mathbb{R}^N)$ whose support,

$$\text{supp}(T) := \bigcap \{ Z \subset \mathbb{R}^n \mid \omega \in \Omega_c^k(\mathbb{R}^N \setminus Z) \implies \langle T, \omega \rangle = 0 \}, \qquad (3.42)$$

is included in X. We denote by $D_k(X)$ the set of currents on X.

Definition 3.60 Let M be a smooth compact SA manifold. A *stratification* of M is a finite partition $M = \bigcup S$ of M such that any $S \in \mathcal{S}$ is a connected smooth submanifold whose closure $\bar{S} = S \cup \bigcup_i T_i$ is the union of S together with elements $T_i \in \mathcal{S}$ of dimension $\dim T_i < \dim S$.

Definition 3.61 Let M be a compact oriented SA manifold of dimension k and let $f : M \to \mathbb{R}^N$ be an SA application. There is a stratification $M = \bigcup S$ such that $f|_S$ is a trivial fiber bundle for any $S \in \mathcal{S}$. Let $S_1, \ldots, S_l$ be the strata such that $f|_{S_i}$ is of rank k. We define a current $f_*[\![M]\!]$ by:

$$\langle f_*[\![M]\!], \omega \rangle := \sum_{i=1}^{l} \int_{S_i} f^* \omega. \qquad (3.43)$$

Definition 3.62 An *SA chain* of degree k on an SA set X is a current of the form $f_*[\![M]\!]$ where $\dim M = k$. We denote by $C_k^{\text{SA}}(X)$ the set of all these currents.

Proposition 3.63 *The set $C_k^{\text{SA}}(X)$ is a subgroup of $D_k(X)$. We have $C_i^{\text{SA}}(X) = 0$ for $i > \dim X$ and $d(f_*[\![M]\!]) = f_*[\![\partial M]\!]$. The collection $C_*^{\text{SA}}(X)$ forms a chain complex, functorial in X with respect to SA applications. There is a canonical natural transformation $\times : C_*^{\text{SA}}(X) \otimes C_*^{\text{SA}}(Y) \to C_*^{\text{SA}}(X \times Y)$ that satisfies the Leibniz formula.*

Definition 3.64 Let X be an SA set and $f_0, \ldots, f_k : X \to \mathbb{R}$ be SA maps. We define an SA cochain $\lambda(f_0; f_1, \ldots, f_k) \in C_{\text{SA}}^k(X) :- C_k^{\text{SA}}(X)^\vee$ by:

$$\forall \gamma \in C_k^{\text{SA}}(X), \ \langle \lambda(f_0; f_1, \ldots, f_k), \gamma \rangle := \langle (f_0, \ldots, f_k)_* \gamma, \ x_0 dx_1 \ldots dx_k \rangle. \tag{3.44}$$

The *minimal forms* on X are the elements of the subgroup $\Omega_{\min}^k(X) \subset C_{\text{SA}}^k(X)$ generated by all the elements of the form $\lambda(f_0; f_1, \ldots, f_k)$.

One should think of the element $\lambda(f_0; f_1, \ldots, f_k)$ as of the form $f_0 \cdot df_1 \wedge \cdots \wedge df_k$.

Proposition 3.65 *Minimal forms define a sub-complex of $C_{\text{SA}}^*(X)$. We have:*

$$d\lambda(f_0; f_1, \ldots, f_k) = \lambda(1; f_0, \ldots, f_k). \tag{3.45}$$

There is a family of maps $\times : \Omega^k_{\min}(X) \otimes \Omega^l_{\min}(Y) \to \Omega^{k+l}_{\min}(X \times Y)$ *which induce a CDGA structure on* $\Omega^*_{\min}(X)$ *using the diagonal* $\Delta : X \to X \times X, x \mapsto (x, x)$:

$$\lambda_1 \cdot \lambda_2 := \Delta^*(\lambda_1 \lambda_2). \tag{3.46}$$

Minimal forms do not define a model of X in general, as the next example shows.

Example 3.66 Consider the space $X = [1, 2]$ and the minimal 1-form $dt/t = \lambda(f_0; f_1) \in \Omega^1_{\min}([1, 2])$ where $f_0(t) = 1/t$ and $f_1(t) = t$. That form is of course closed, as dim $X = 1$. As we have $H^1([1, 2]) = 0$, this form would have to be exact. But it is not the coboundary of any form of degree 0, as the logarithm is not an SA map.

To solve this problem, formal integrals along fibers of SA bundles must be introduced. More generally, one must be able to integrate along any continuous family of chains.

Definition 3.67 Let $f : Y \to X$ an SA application. A *strongly continuous family of chains* of degree l on over X is a map $\Phi : X \to C_l(Y)$ such that there exists:

- a finite SA stratification $X = \bigcup_{\alpha \in I} S_\alpha$;
- closed manifolds oriented SA F_α of dimension l;
- SA maps $g_\alpha : \bar{S}_\alpha \times F_\alpha \to Y$ such that:

  - the following diagram commutes:

$$
\begin{array}{ccc}
\bar{S}_\alpha \times F_\alpha & \xrightarrow{\ g_\alpha\ } & Y \\
\Big\downarrow{\scriptstyle p_{\bar{S}_\alpha}} & & \Big\downarrow{\scriptstyle f} \\
\bar{S}_\alpha & \lhook\joinrel\longrightarrow & X,
\end{array}
\tag{3.47}
$$

  - for all α, for all $x \in \bar{S}_\alpha$, one has $\Phi(x) = (g_\alpha)_*[\![\{x\} \times F_\alpha]\!]$.

Remark 3.68 As explained in [HLTV11 Remark 5.14], it is crucial to consider the closure of S_α in the previous definition, rather than just S_α, as it represents a kind of continuity condition. The definition given by Kontsevich and Soibelman [KS00] was weaker.

Example 3.69 Let $\pi : E \to B$ be an SA fiber bundle of rank l. One can define a strongly continuous family of chains $\Phi : B \to C_l(E)$ by taking $\Phi(b) := [\![\pi^{-1}(b)]\!]$.

Proposition 3.70 Let $\gamma \in C^{SA}_k(X)$ be an SA chain and $\Phi : X \to C_l(Y)$ be a *strongly continuous family of chains, with the notation above. We can refine the stratification to have* $\gamma = \sum n_\alpha \cdot [\![\bar{S}_\alpha]\!]$. *We can then define a new SA chain* $\gamma \ltimes \Phi \in C^{SA}_{k+l}(Y)$ *by:*

$$\gamma \ltimes \Phi := \sum_\alpha n_\alpha (g_\alpha)_* [\![\bar{S}_\alpha \times F_\alpha]\!]. \tag{3.48}$$

Definition 3.71 Let $\lambda \in \Omega_{\min}^{k+l}(Y)$ be a minimal form on Y and let $\Phi : X \to C_l^{SA}(Y)$ be a strongly continuous family of chains. We define an SA cochain $\int_\Phi \lambda$ by:

$$\left\langle \int_\Phi \lambda, \gamma \right\rangle := \langle \lambda, \gamma \ltimes \Phi \rangle. \tag{3.49}$$

The *piecewise semi-algebraic (PA) forms* of degree k over X are the cochains of the form $\int_\Phi \lambda$. One notes $\Omega_{PA}^k(X) \subset C_{SA}^k$ the submodule that they span.

Remark 3.72 A minimal form is in particular a PA form. Indeed, if $\omega \in \Omega_{\min}^k(X)$ is a minimal form, one can take $\Phi : X \to C_0(X)$, $x \mapsto [[\{x\}]]$. Then $\int_\Phi \omega = \omega$.

Note that the proof of the second part of the following proposition is nontrivial.

Proposition 3.73 *The collection $\Omega_{PA}^*(X)$ is a sub-complex of $C_{SA}^*(X)$ which vanishes in degree $> \dim X$. There is a multiplication on $\Omega_{PA}^*(X)$ which extends that of $\Omega_{\min}^*(X)$ and makes it a CDGA.*

One of the main results of [HLTV11] is the following:

Theorem 3.74 ([HLTV11, Theorems 6.1, 7.1]) *There is a zigzag of natural transformations, natural in SA sets X and SA maps:*

$$\Omega_{PA}^*(X) \leftarrow \cdot \to \Omega_{PL}^*(X) \otimes_\mathbb{Q} \mathbb{R} \tag{3.50}$$

which are quasi-isomorphisms if X is a compact SA set. This zigzag is moreover compatible with Künneth morphisms.

Remark 3.75 Hardt et al. [HLTV11, Section 9.1] conjecture that the result remains true if X is not compact.

Example 3.76 The form $dt/t \in \Omega_{\min}^1([1, 2])$ of Example 3.66 is the coboundary of a PA form. Let $\lambda = \lambda(f_0; f_1) \in \Omega_{\min}^1([1, 2]^2)$ be the minimal form associated to $f_0(s, t) = 1/s \cdot \chi_{t<s}$ and $f_1(s, t) = s$, where $\chi_{t<s}$ is the indicator function. Consider the strongly continuous family of chains $\Phi : [1, 2] \to C_1^{SA}([1, 2]^2)$ associated to the projection on the second factor. Then we can define:

$$\log t := \int_\Phi \lambda = \int_1^2 \chi_{s<t} ds \in \Omega_{PA}^0([1, 2]), \tag{3.51}$$

for $t \in [1, 2]$ and we have $d(\log t) = dt/t$.

The proof of Theorem 3.74 (that we are not going to detail as it is highly technical) essentially involves three ingredients:

- the Poincaré lemma: $\Omega_{PA}^*(\Delta^n)$ is acyclic for fixed n;

- the simplicial sets $\Omega_{PA}^k(\Delta^\bullet)$ are "extendable" for fixed k, i.e., for any subset $I \subset \{0, \ldots, n\}$, for any collection $\{\beta_i \in \Omega_{PA}^k(\Delta^{n-1})\}_{i \in I}$ of forms satisfying $d_i \beta_j = d_{j-1}\beta_i$, there is a k form $\beta \in \Omega_{PA}^k(\Delta^n)$ checking $d_i \beta = \beta_i$.
- the Mayer–Vietoris property, i.e., $\Omega_{PA}^*(U \cup V)$ can be computed from $\Omega_{PA}^*(U)$, $\Omega_{PA}^*(V)$, and $\Omega_{PA}^*(U \cap V)$.

Using these three properties, the proof of the theorem follows from categorical reasons.

Let us now turn to integration along fibers of PA forms. Recall that an SA fiber bundle $\pi : E \to B$ of rank l defines a strongly continuous family of chains $\Phi : B \to C_l(E)$.

Definition 3.77 Given a minimal form $\omega \in \Omega_{\min}^{k+l}(E)$ and an SA fiber bundle $\pi : E \to B$ of rank l, one defines the integral of ω along the fibers of π by:

$$\pi_*(\omega) = \int_{\pi : E \to B} \omega := \int_\Phi \omega \in \Omega_{PA}^k(B). \tag{3.52}$$

This procedure has many properties similar to those of the conventional integrals along fibers. The property that will interest us the most is Stokes' formula. We recall that $\pi^\partial : E^\partial \to B$ is the SA fiber bundle of rank $l - 1$ given by the fiberwise boundary of E, see Definition 3.41

Proposition 3.78 ([HLTV11, Proposition 8.12]) *Let $\pi : E \to B$ be an oriented SA bundle of rank l and let $\omega \in \Omega_{\min}^{k+l}(E)$ be a minimal form. The Stokes formula holds:*

$$d(\pi_*(\omega)) = \pi_*(d\omega) + (-1)^{\deg \omega - l} \pi_*^\partial(\omega|_{E^\partial}). \tag{3.53}$$

Let us also record the following facts, which will be useful in what follows.

Proposition 3.79 ([HLTV11, Proposition 8.11]) *Let $\pi : E \to B$ be an oriented SA fiber bundle of rank l. Suppose that $E = \bigcup_{i \in I} E_i$ is a union of SA subsets such that $\pi|_{E_i}$ remains an SA bundle of rank l and that $\dim \pi|_{E_i}^{-1}(x) \cap \pi|_{E_j}^{-1}(x) < l$ for all $i \neq j$ and all $x \in B$. Then one has $\pi_*(\omega) = \sum_{i \in I} (\pi|_{E_i})_*(\omega|_{E_i})$.*

Proposition 3.80 *Let M be a compact SA manifold with boundary. Then the restriction map $\Omega_{PA}^*(M) \to \Omega_{PA}^*(\partial D)$ is surjective.*

Proof Choose an SA function $f : [0, 1) \to [0, 1]$ such that $f(0) = 1$ and $f(x) = 0$ for $x > 1/2$ (for example, a piecewise affine function). Let $\partial M \subset V \subset M$ be an SA tubular neighborhood of ∂M, that is, $V \cong \partial M \cong [0, 1)$ and ∂M is $\partial M \times \{0\} \subset V$. There exists a retraction $\rho : V \to \partial M$ of the inclusion $\partial M \subset V$. Then given any form $\omega \in \Omega_{PA}^*(\partial M)$, the form $f \cdot \rho^*\omega$ can be extended by zero outside of V. This extension by zero is a PA form that restricts to ω on ∂M. $\qquad\square$

3.4 Graph Complexes

In this section, we introduce one of the main constructions of the chapter, the graph complexes. These graph complexes are used to bridge the Lambrechts–Stanley model to the forms on the configuration spaces. These graph complexes depend on analytic data called the partition function (see Sect. 3.4.5). A key step of the proof will be the fact that this partition function is trivial up to homotopy, which we show in Sect. 3.4.6.

As in Sect. 3.2, we will consider that our CDGA collections are no longer indexed by integers but by any finite set. In particular, we have $G_A(r) = G_A(\{1, \ldots, r\})$.

3.4.1 Informal Idea

Let M be a simply connected closed SA manifold and let A be a Poincaré duality model of A. Our objective is to show that the CDGAs $G_A(U)$ and $\Omega_{PA}^*(FM_M(U))$ are quasi-isomorphic. Finding a direct quasi-isomorphism would be a miracle: since the CDGA $G_A(U)$ has many relations, one would have to find PA forms on $FM_M(U)$ that strictly satisfy these relations.

As often in homological algebra, we will define a *resolution* of $G_A(U)$, i.e., a quasi-free CDGA which is quasi-isomorphic to $G_A(U)$. All the complexity of the relations is then transferred into the differential. Finding a morphism from this resolution to $\Omega_{PA}^*(FM_M(U))$ then only requires finding forms that are compatible with the differential. This will essentially consist in finding forms that satisfy the relations up to homotopy.

Let us now explain the general philosophy behind the graph complexes that we use. The relations in the CDGA $G_A(U)$ are of three types:

1. the relations that exist in A;
2. the Arnold relations, and especially the three-term relation;
3. the symmetry relation $p_i^*(a)\omega_{ij} = p_j^*(a)\omega_{ij}$.

The first type of relation depends on the manifold M. Given a simply connected manifold, we can find a quasi-free resolution $R \to \Omega_{PA}^*(M)$ and apply the procedure of Lambrechts and Stanley [LS08b] to find a quasi-isomorphism $R \to A$, where A is a Poincaré duality CDGA.

For the second type of relations, we will use an idea due to Kontsevich. Recall the graphical description of $G_A(U)$ found after Definition 3.15. Kontsevich's idea is to replace the three-term relation $\omega_{ij}\omega_{jk} + \omega_{ki} + \omega_{ki}\omega_{ij} = 0$ by a differential in a graph complex. This graph complex is a vector space spanned by graphs whose vertices are of two types:

1. The first kind of vertices will be called "external". These external vertices are the same that already appeared in the graphical description of $G_A(U)$ after

Definition 3.15. These external vertices are equipped with a bijection with the fixed finite set U.

2. The second kind of vertices will be called "internal". These vertices, which will be drawn as filled black circles in pictures, are unnumbered. More formally, we are first going to consider graph with numbered additional vertices, and we are then going to take the coinvariants of the set of linear combination of such graphs under the action of the symmetric group which renumbers the additional ("internal") vertices. There can be an arbitrary number of internal vertices in a graph (including zero).

The differential consists in contracting the edges incident to the internal vertices. We can then graphically represent the three-term relation by the following picture.

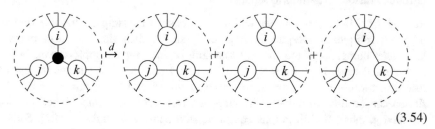

$$(3.54)$$

The four graphs that appear in this formula are identical outside of the dashed circle. The graphs all have (at least) three external vertices indexed by i, j, and k (and possibly other external vertices outside of the circle). The graph on the left-hand side of the equation has an extra internal vertex compared to the other three (the one drawn in black). The graphs one the right-hand side are three of the summands of the differential. These summands are obtained by contracting the edges incident to the internal vertex and reconnecting the other edges that were incident to the internal vertex to the other extremity of the contracted edge.

The connection between this graph complex and configuration spaces is the following. A graph with external vertices U and internal vertices I corresponds to a form on $\mathsf{FM}_M(U)$ obtained by taking a form on $\mathsf{FM}_M(U \sqcup I)$ and integrating it along the fibers of the projection $\mathsf{FM}_M(U \sqcup I) \to \mathsf{FM}_M(U)$. The internal vertex in (3.54) corresponds to an extra temporary point in a possible configuration on M. The integral along the fibers of the projection is an "average" over all possible positions of this extra point. When the Stokes formula is applied, one considers the fiberwise boundary. This boundary has three faces of maximal dimension: when the fourth point becomes infinitely close to one of the other three points with respect to the other two. These three faces correspond exactly to the three terms of Arnold's relation.

The last type of relation, symmetry, is managed in a similar way to Kontsevich's idea. More precisely, this symmetric relation becomes a boundary as shown in the following figure which depicts how the symmetry relation is obtained as part of a coboundary:

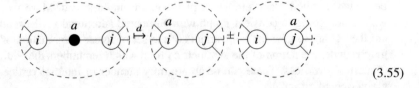

$$(3.55)$$

Of course, in the complete graph complex, these two differentials are mixed with the cutting part of the differential that reflects $d\omega_{ij} = \Delta_{ij}$. This will lead us to introduce various filtrations to separate them.

Remark 3.81 In this volume, we will mainly consider the simplest possible version of graph complexes. Despite the apparently simple definition, graph complexes hide deep complexity, and the full structure of the graph complex, even in the undecorated case, is still largely a mystery and constitutes a current subject of research. Moreover, there are many variants: ribbon graphs (where the twist of a ribbon can be used to model a framing, see Definition 5.53), undirected graphs, oriented graphs [Živ20], graphs equipped with sources or sinks [Živ21], Swiss-Cheese-type graphs (see Sect. 5.4.3 and [Idr20; Wil15]), graphs with external legs called "hairs" (see e.g. [AT15; FTW17]), and so on.

3.4.2 Definition of the Unreduced Graph Complex

We now fix a zigzag of quasi-isomorphisms

$$A \leftarrow R \rightarrow \Omega^*_{\mathrm{PA}}(M), \tag{3.56}$$

where A is a Poincaré duality CDGA and R is a quasi-free CDGA generated in degrees ≥ 2. This zigzag exists thanks to a theorem of Lambrechts and Stanley [LS08b].

We also fix a cocycle $\Delta_R \in R \otimes R$ of degree n which is (anti)symmetric and is mapped to Δ_A by the application $R \rightarrow A$. Such a cocycle exists by elementary arguments. It should be noted, however, that it is necessary to assume that $A = H^*(M)$ if $n \leq 6$ (any simply connected manifold of dimension ≤ 6 is formal), or that the above zigzag verifies a compatibility relation between the integration $\int_M : \Omega^n_{\mathrm{PA}}(M) \rightarrow \mathbb{R}$ and the Poincaré duality augmentation $\varepsilon : A^n \rightarrow \mathbb{R}$. This means that a priori, we cannot use any Poincaré duality model of M. However, we will prove at the end of the chapter that if B is another Poincaré duality model, then $\mathsf{G}_B \simeq \mathsf{G}_A$ (see Proposition 3.125).

We are going to build a first graph complex, which will however have the wrong homotopy type (see Sect. 3.4.3). This issue be solved in Sect. 3.4.5. We are first going to give a graphical description of the graph complex. We give a more algebraic description in terms of operadic twisting in Remark 3.4.2.1. This issue will be solved in Sect. 3.4.5.

Definition 3.82 The graph complex $\mathsf{Graphs}'_R(U)$ is the vector space generated by equivalence classes of graphs of the following type.

- There are no double edges or loops. More precisely, a graph containing a double edge or a loop is identified with 0;
- The vertices of the graph are split into two different categories.

    - The vertices of the first category are called "external". The set of external vertices is equipped with a bijection with U. In pictures, they will be drawn as circles containing their index $u \in U$.
    - The rest of the vertices are said to be "internal". They are drawn as black circles in pictures. Informally, internal vertices are unnumbered, which means that they cannot be distinguished from one another. Strictly speaking, one first consider a version where internal vertices are numbered, then the quotient by the action of the symmetric group permuting internal vertices, up to a sign.

- Each vertex—internal or external—is decorated by an element of R;
- If n is odd, then each edge is equipped with an orientation. However, a graph is identified with the opposite of the graph where the direction of an edge has been swapped.
- If n is even, then the set of all edges is equipped with an order. However, a graph is identified with the opposite of a graph that differs from it by an edge transposition.

See Fig. 3.5 for an example. Thanks to the last two conditions, we can forego the orientations or the order of the edges in pictures, at the cost of undetermined signs. The degree of a graph is calculated by adding the degrees of all the decorations, adding $n - 1$ for each edge, and subtracting n for each internal vertex.

If $\Gamma, \Gamma' \in \mathsf{Graphs}'_R(U)$ are two such graphs, then their product $\Gamma \cdot \Gamma'$ is the graph obtained by gluing Γ and Γ' along their external vertices. The decorations of the

Fig. 3.5 Example of a graph in $\mathsf{Graphs}'_R(\{u, v, w\})$ where $x, y, z, a, b \in R$

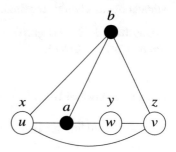

corresponding external vertices are multiplied during the operation. For example, we have the following equation in $\mathsf{Graphs}'_R(\{i, j, k\})$:

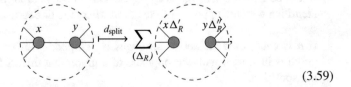

$$\tag{3.57}$$

Finally, the differential

$$d = d_R + d_{\text{split}} + d_{\text{contr}} \tag{3.58}$$

is the sum of three terms:

- the internal differential d_R of R, which acts on each decoration like a derivation;
- a "splitting" part d_{split}, which is the sum of all possible ways to cut an edge and multiply the end decorations by Δ_R as in the following picture (the gray vertices can be internal or external):

$$\tag{3.59}$$

- a "contracting" part d_{contr}, which is the sum of all possible ways to contract an edge incident to an internal vertex by multiplying the decorations (if the two vertices incident to the edge are internal, the result is an internal vertex, otherwise it is an external vertex of the same number as the only external vertex incident to the edge), see Eq. (3.54).

Checking the following proposition can be done easily "by hand" (but one has to be careful of the signs). One can also use operadic twisting theory to reduce the amount of calculations required.

Proposition 3.83 *This graph complex* $\mathsf{Graphs}'_R(U)$ *with this differential and this product forms a CDGA.*

3.4.2.1 A Formal Definition Through Operadic Twisting

This idea of adding internal vertices to solve a three-term relation of the previous type (which can be seen as the dual of a Jacobi relation) has been formalized and generalized by the notion of "operadic twisting". We refer to Willwacher [Wil14, Appendix I] and Dolgushev and Willwacher [DW15] for more precise definitions.

Note that we are going to remain slightly vague about signs below. In order to have more precise definitions, one needs the theory of operads, which is the subject of Chap. 5.

Let us now quickly outline how operadic twisting works in the case of our graph complex. We first define the untwisted R-decorated graph complex $\mathsf{Gra}_R(V)$ (for some finite set V) by:

$$\mathsf{Gra}_R^{\circlearrowleft}(V) := \left(R^{\otimes V} \otimes S(e_{ij})_{i,j \in V}/(e_{ji} - (-1)^n e_{ij}), d(e_{ij}) = (p_i^* \cdot p_j^*)(\Delta_R)\right). \tag{3.60}$$

The generators e_{ij} (of degree $n - 1$) correspond to the edges. The factor $R^{\otimes V}$ corresponds to the decoration. A word of the form $\bigotimes_{v \in V} r_v \otimes e_{i_1 j_1} \ldots e_{i_s j_s}$ can be interpreted as a graph on $|V|$ numbered vertices, where the vertex labeled by $v \in V$ is decorated by r_v, and there is an edge between the vertices i and j for every term e_{ij} that appears in the monomial.

Example 3.84 Let us consider Eq. (3.57). This equation can be rewritten algebraically as:

$$(x \otimes y \otimes z) \otimes e_{jk} \cdot (x' \otimes y' \otimes z') \otimes e_{ik} = (xx' \otimes yy' \otimes zz') \otimes e_{jk}e_{ik}. \tag{3.61}$$

Remark 3.85 The generators of $\mathsf{Gra}_R(V)$ correspond to graphs whose vertices are all external and where tadpoles (elements containing a term of the form e_{ii}) and double edges (elements containing a term of the form e_{ij}^2) are allowed. This digression through the larger graph complex that contains tadpoles is necessary to be able to apply directly the theory of operadic twisting when the Euler characteristic $\chi(M)$ is nonzero. When $\chi(M) \neq 0$, there is no action of the graphs operad (see Proposition 5.78). In the end, we will consider a subcomplex of graphs without tadpoles or double edges. Note moreover that depending on the parity of n, either tadpoles our double edges are automatically killed:

- If n is even, then $\deg e_{ij} = n - 1$ is odd and therefore $e_{ij}^2 = -e_{ij}^2$ must vanish, i.e., there are no double edges;
- If n is odd, then the relation $e_{ji} = -e_{ij}$ applied to $i = j$ yields $e_{ii} = 0$, i.e., there are not tadpoles.

One then considers the twisted version, defined for a finite set U by:

$$\mathsf{Tw}\,\mathsf{Gra}_R^{\circlearrowleft}(U) := \prod_{i \geq 0}\left(\mathsf{Gra}_R^{\circlearrowleft}(U \sqcup \{1, \ldots, i\}) \otimes \mathbb{R}[n]^{\otimes i}\right)_{\mathfrak{S}_i}, \tag{3.62}$$

where the symmetric group acts on $\mathsf{Gra}_R^{\circlearrowleft}(U \sqcup \{1, \ldots, i\})$ by renumbering, and on $\mathbb{R}[n]^{\otimes i} \cong \mathbb{R}[ni]$ by the Koszul rule of signs. The differential of Eq. (3.58) has three terms, just like before:

1. the internal differential of R, which acts on decorations;
2. the differential induced by that of $\mathsf{Gra}_R^{\circlearrowleft}$, which acts on edges ($de_{ij} = \Delta_{ij}$);

3. a third term which is given by the sum of all the ways of removing a monomial of the type e_{uv} where at least one of the two factors is internal (i.e., an element of $\{1, \ldots, i\}$) and renumbering all the other indices so that the two extremities become only one vertex.

Example 3.86 Consider the graph of Fig. 3.5. We can temporarily index one of the internal (black) vertices by 1 and the other by 2 before taking coinvariants. This graph can be seen as the equivalence class of the element:

$$(x \otimes y \otimes z \otimes a \otimes b) \otimes e_{u1}e_{u2}e_{uv}e_{12}e_{1w}e_{vw}e_{w2} \in \mathsf{Gra}_R^{\circlearrowleft}(\{u, v, w, 1, 2\}). \qquad (3.63)$$

One then give a degree of $-n$ to the black vertices (which corresponds to the $\mathbb{R}[n]^{\otimes 2}$ factors) and takes the coinvariants under the action of $\mathfrak{S}_2$. The third term of the differential (edge contraction) will have five terms. These five terms correspond to the five edges incident to the black vertices.

Finally, in order to obtain $\mathsf{Graphs}'_R(U)$, we simply take the subcomplex of $\mathrm{Tw}\,\mathsf{Gra}_R^{\circlearrowleft}(U)$ spanned by graphs which have neither tadpoles nor double edges. It is easy to see that this is a subcomplex using the observation at the end of Remark 3.85.

3.4.3   An Issue with the Homotopy Type

This graph complex unfortunately has the wrong homotopy type. The problem comes from the internal components of the graphs, i.e., the connected components consisting exclusively of internal vertices.

Example 3.87 In the undecorated case (i.e., $R = \mathbb{R}$) and for $n = 2$, the cohomology of the internal part of the graph complex contains the Grothendieck–Teichmüller Lie algebra $\mathfrak{grt}_1$ [Wil14], which is infinite-dimensional. This is very different from the cohomology of $\mathsf{Conf}_{\mathbb{R}^2}(0)$, which is merely a singleton.

Unfortunately, one cannot simply mod out these internal components, as the following example shows:

$$\left(\underbrace{\textcircled{1}}\!\!-\!\!\bullet\right) \overset{x}{} \overset{d}{\longmapsto} \left(\underbrace{\textcircled{1}}\!\!-\!\!\bullet\right) \overset{d_R x}{} \pm \sum_{(\Delta_R)} \left(\underbrace{\textcircled{1}}^{\Delta'_R} \quad \bullet^{x\Delta''_R}\right) \pm \textcircled{1}^{\,x}. \qquad (3.64)$$

If one killed all graphs with internal components, then all graphs with a single external vertex of $\mathsf{Graphs}'_R(1)$ would become zero in cohomology. But these graphs correspond to the elements of R, thus to the cohomology classes of M. It would of course be absurd to mod out by all these classes.

However, if we remember Lemma 3.12, we see that it would be enough to identify an isolated internal vertex decorated by $x \in R$ with the number $\varepsilon(x)$ to have coherent relations. All that remains then is to deal with the internal components containing at least two vertices. One could choose to simply kill them, as they cannot cause an issue similar to the one of Eq. (3.64). If we do so, then we can build a quotient map $\mathsf{Graphs}'_R(U)/\sim \to \mathsf{G}_A(U)$ and prove in a combinatorial way that it is a quasi-isomorphism (Sect. 3.5). However, it will then not be possible to build a map $\mathsf{Graphs}'_R(U)/\sim \to \Omega^*_{\mathrm{PA}}(\mathsf{FM}_M(U))$ that preserves the differential. Indeed, the splitting part of the differential can create graphs with internal components, and the Stokes formula forces to identify them with integrals on $\mathsf{FM}_M(U)$. We will explain in the next sections a way around this issue.

3.4.4 Propagator

We will now define a morphism $\omega : \mathsf{Graphs}'_R(U) \to \Omega^*_{\mathrm{PA}}(\mathsf{FM}_M(U))$. As announced earlier, the idea will be to consider the external vertices as "fixed" points in a configuration, while the internal vertices will correspond to moving points that will be averaged over all possible positions.

We will simply use the fixed morphism $R \to \Omega^*_{\mathrm{PA}}(M)$ fixed at the beginning to know where to send the vertices decorations. It then remains to know where to send the edges. We must thus find a form $\varphi \in \Omega^{n-1}_{\mathrm{PA}}(\mathsf{FM}_M(2))$ which must satisfy several conditions so that the morphism $\mathsf{Graphs}'_R(U) \to \Omega^*_{\mathrm{PA}}(\mathsf{FM}_M(U))$ is well defined. In mathematical physics, such a form φ is called a propagator. One finds the conditions that φ must check by reasoning about the properties of the edges in the graph complex, and by trying to derive the compatibility $\omega d = d\omega$ from Stokes' formula.

Let us note that the two fiber bundles $p_1, p_2 : \mathsf{FM}_M(2) \to M$ restrict to the same bundle $p : \partial\mathsf{FM}_M(2) \to M$. This fiber bundle p is the sphere bundle of rank $n-1$ associated with the tangent bundle of M (see Example 3.49).

Proposition 3.88 ([CW16; CM10]) *There exists a form $\varphi \in \Omega^{n-1}_{\mathrm{PA}}(\mathsf{FM}_M(2))$, called the propagator, satisfying the following conditions.*

- *The form is (anti)symmetric: if σ is the automorphism of $\mathsf{FM}_M(2)$ which exchanges the two points, then $\sigma^*\varphi = (-1)^n\varphi$.*
- *The differential of φ is the diagonal class: $d\varphi = (p_1, p_2)^*(\Delta_M)$, where Δ_M is the image of Δ_R by the fixed map $R \to \Omega^*_{\mathrm{PA}}(M)$ and $(p_1, p_2) : \mathsf{FM}_M(2) \to M \times M$ is the product of the two projections.*
- *The restriction of φ to $\partial\mathsf{FM}_M(2)$ is a global angular form: on each fiber of $p : \partial\mathsf{FM}_M(2) \to M$, the form φ restricts to a volume form.*

Proof The existence of such a φ in our framework has been proved by Campos and Willwacher [CW16], based on calculations previously made by Cattaneo and Mnëv [CM10]. To construct φ, one starts by choosing a global angular form $\psi \in \Omega^{n-1}_{\mathrm{PA}}(\partial\mathsf{FM}_M(2))$, which exists in general (see for example Bott and Tu [BT82]). The standard proof is in the smooth setting, but it can be easily adapted to the SA setting. One can furthermore choose ψ such that $d\psi$ is basic, i.e., it is the pullback of a form on M (namely, the Euler class of M). By considering the (anti)symmetrization, we can assume that $\sigma^*\psi = (-1)^n\psi$. We can find a tubular neighborhood $\rho : T \twoheadrightarrow \partial\mathsf{FM}_M(2)$ inside $\mathsf{FM}_M(2)$, which allows us to extend ψ into $\rho^*\psi$. Taking an approximation of the indicator function of $\partial\mathsf{FM}_M(2)$, we can extend $\rho^*\psi$ to a form $\psi' \in \Omega^{n-1}(\mathsf{FM}_M(2))$ which is equal to ψ on $\partial\mathsf{FM}_M(2)$, and which vanishes outside T. As $d\psi'|_{\partial\mathsf{FM}_M(2)}$ is basic, the form $d\psi'$ is the pullback of a form $\alpha \in \Omega^n_{\mathrm{PA}}(M \times M)$ which is closed but not (necessarily) exact. We can compute that $[\alpha] \in H^n(M \times M)$ is the diagonal class. Indeed, if β is any other closed form, then:

$$\int_{M \times M} \alpha \wedge \beta = \int_{\mathsf{FM}_M(2)} d\psi' \wedge (p_1, p_2)^*(\beta)$$

$$= \int_{\partial\mathsf{FM}_M(2)} \psi \wedge (p_1, p_2)^*(\beta)|_{\partial M} = \int_{\Delta_M} \beta_{\Delta_M}, \qquad (3.65)$$

where $\Delta_M = \{(x, x) \in M^2\} \cong M$. There is therefore a form γ such that $\alpha - \Delta_M = d\gamma$ on $M \times M$. We can then define $\varphi := \psi' - (p_1, p_2)^*\gamma$. We then have $\varphi = (p_1, p_2)^*(\Delta_M)$. As $p^*\gamma|_{\partial\mathsf{FM}_M(2)}$ is basic, the form φ remains (anti)symmetric and its restriction to the boundary remains a global angular form.   □

We can now define the morphism $\omega : \mathsf{Graphs}'_R(U) \to \Omega^*_{\mathrm{PA}}(\mathsf{FM}_M(U))$.

Definition 3.89 Let $\Gamma \in \mathsf{Graphs}'_R(U)$ be a graph whose set of edges is E and whose set of internal vertices is I. We define $\omega'(\Gamma) \in \Omega^*_{\mathrm{PA}}(\mathsf{FM}_M(U \sqcup I))$ by:

$$\omega'(\Gamma) := \bigwedge_{v \in U \sqcup I} p_v^*(\alpha_v) \wedge \bigwedge_{e \in E} p_e^*(\varphi). \qquad (3.66)$$

In this formula, $\alpha_v \in R$ is the vertex decoration of $v \in U \sqcup I$, the map $p_v : \mathsf{FM}_M(U \sqcup I) \to M$ is the projection that forgets all points except v, and the map $p_e : \mathsf{FM}_M(U) \to \mathsf{FM}_M(2)$ is the projection that forgets all points except the extremities of the edge e. Let $\pi : \mathsf{FM}_M(U \sqcup I) \to \mathsf{FM}_M(U)$ the canonical projection. Then we also define:

$$\omega(\Gamma) := \pi_*(\omega'(\Gamma)) = \int_{\mathsf{FM}_M(U \sqcup I) \to \mathsf{FM}_M(U)} \omega'(\Gamma). \qquad (3.67)$$

Remark 3.90 There is something to be said at this point. The form $\omega'(\Gamma)$ is not minimal in general. We thus cannot compute, a priori, its integral along the fibers of π. Campos and Willwacher [CW16] have defined a sub-CDGA of $\Omega^*_{\mathrm{PA}}(-)$ made up of "trivial" forms, i.e., the forms that are obtained by integrating minimal forms along trivial fiber bundles. They showed that this sub-CDGA is quasi-isomorphic to $\Omega^*_{\mathrm{PA}}(-)$ and that a trivial form can be integrated along the fibers of any fiber bundle.

They also showed that one can make $R \to \Omega^*_{\mathrm{PA}}(M)$ factors through the trivial forms and that one can choose a propagator that is a trivial form.

Proposition 3.91 *The map $\omega : \mathsf{Graphs}'_R(U) \to \mathsf{FM}_M(U)$ is a CDGA morphism.*

Proof It is easily verified that ω' is compatible with the identifications (permutations of edges or orientation reversals) that define $\mathsf{Graphs}'_R(U)$.

Let $\Gamma, \Gamma' \in \mathsf{Graphs}'_R(U)$ be two graphs, with respective sets of internal vertices I and I'. Let us compare $\omega(\Gamma' \cdot \Gamma')$ with $\omega(\Gamma) \wedge \omega(\Gamma')$. There is a commutative square:

$$
\begin{array}{ccc}
\mathsf{FM}_M(U \sqcup I \sqcup I') & \xrightarrow{\ q\ } & \mathsf{FM}_M(U \sqcup I) \\[2mm]
\Big\downarrow{\scriptstyle q'} & \overset{\pi}{\searrow} & \Big\downarrow{\scriptstyle p} \\[2mm]
\mathsf{FM}_M(U \sqcup I') & \xrightarrow{\ p'\ } & \mathsf{FM}_M(U)
\end{array}
\tag{3.68}
$$

This square is unfortunately not Cartesian, as $\mathsf{FM}_M(U \sqcup I \sqcup I')$ is not the fiber product $P = \mathsf{FM}_M(U \sqcup I) \times_{\mathsf{FM}_M(U)} \mathsf{FM}_M(U \sqcup I')$ of the other three spaces. We can interpret P as a "virtual" configuration space, where the points of $U \sqcup I$ form a configuration, those of $U \sqcup I'$ form a configuration, but the points of I and I' can collide. We have an induced map $\rho : \mathsf{FM}_M(U \sqcup I \sqcup I') \to P$ which is a morphism of SA bundles above $\mathsf{FM}_M(U)$. This map is of degree 1 on the fibers (i.e., it sends the fundamental class to the fundamental class). As a consequence, if $\alpha \in \Omega^*_{\mathrm{PA}}(P)$ is any form, the integral along the fibers of $\rho^*(\alpha)$ is equal to the integral along the fibers of α. The form $\omega'(\Gamma \cdot \Gamma')$ is of the type $\rho^*(q^*(\alpha) \wedge q'^*(\alpha'))$, where $\alpha = \omega'(\Gamma)$ and $\alpha' = \omega'(\Gamma')$. It follows that the integral $\omega(\Gamma \cdot \Gamma')$ separates into two factors $p_*(\alpha) \wedge p'_*(\alpha')$ thanks to a general theorem on integrals along the fibers. These two factors can be identified respectively with $\omega(\Gamma)$ and $\omega(\Gamma')$. This shows that ω is a morphism of algebras.

It then only remains to show that ω is compatible with the differential. This follows from Stokes' formula and from the description of the fiberwise boundary of the projection $\mathsf{FM}_M(U \sqcup I) \to \mathsf{FM}_M(U)$ from Proposition 3.53. By the Stokes formula (Proposition 3.78), we have:

$$
d(\pi_*(\omega'(\Gamma))) = \pi_*(d(\omega'(\Gamma))) \pm \pi^\partial_*(\omega'(\Gamma)|_{\mathsf{FM}^\partial_M(U \sqcup I)}).
\tag{3.69}
$$

The first term corresponds to the internal differential of R (acting on the decorations) and to the splitting part d_{split} (thanks to $d\varphi = \Delta_M$). We can identify the second term to the contracting part of the differential. Let us set $A = U \sqcup I$ and let us recall that the fiberwise boundary can be expressed as (Proposition 3.53):

$$
\mathsf{FM}^\partial_M(A) = \bigcup_{W \in \mathcal{BF}_M(A,U)} \partial_W \mathsf{FM}_M(A),
\tag{3.70}
$$

where $\mathcal{BF}_M(A, U) = \{W \subset A \mid \#W \geq 2 \text{ and } \#(W \cap U) \leq 1\}$. The fiberwise integral $\pi^\partial_*(\omega'(\Gamma)|_{\mathsf{FM}^\partial_M(U \sqcup I)})$ splits as a sum of fiberwise integrals over these facets thanks to Proposition 3.79.

The facet $\partial_W \mathsf{FM}_M(A)$ is given by the set of configurations where the points indexed by W are infinitesimally close to each other with respect to the points that are not in W. This facet is the total space of a fiber bundle of the form $\mathsf{FM}_n(W) \hookrightarrow \partial_W \mathsf{FM}_M(A) \twoheadrightarrow \mathsf{FM}_M(A/W)$. Thanks to the condition $\#(W \cap U) \leq 1$, we can identify A/W with $U \sqcup J$ for a certain subset $J \subset I$ of internal. Let $\Gamma_W \subset \Gamma$ be the full subgraph on the vertices of W. We see that $\int_{\partial_W \mathsf{FM}_M(A) \to \mathsf{FM}_M(U)} \omega'(\Gamma)$ is equal to $c_{\Gamma_W} \omega(\Gamma/\Gamma_W)$, where c_{Γ_W} is either the form given by the integral of $\omega'(\Gamma_W)$ along the fibers of the bundle $\partial_W \mathsf{FM}_M(W) \to \mathsf{FM}_M(U \cap W)$, if $\#(W \cap U) = 1$, or the number given by the integral of $\omega'(\Gamma_W)$ on $\partial_W \mathsf{FM}_M(W)$ otherwise.

We can show that c_{Γ_W} vanishes unless $\#W = 2$ and the two points are connected by a single edge. Indeed, we can show that c_{Γ_W} vanishes if:

- Γ_W is disconnected, by an argument on the dimension and an argument similar to the one that shows that ω is a morphism of algebras;
- Γ_W contains a univalent vertex, unless it is the graph with exactly two vertices (again by a similar argument);
- Γ_W contains a bivalent vertex by a symmetric argument (and also by a similar argument to the one that shows that ω preserves the product);
- or Γ_W contains at least three vertices: in general by a degree counting argument (if e is the number of edges and v the number of vertices, then $e \geq 3v/2$, the form to be integrated is of degree $(n-1)e \geq 3v(n-1)/2 = nv + (n-3)v/2 \geq nv$, and the space $\mathsf{FM}_n(v)$ on which it is integrated is of dimension $nv - n - 1 < nv$) or by an ad-hoc argument due to Kontsevich for $n = 2$.

This leaves only the terms of the form $\pm \Gamma/e$ for an edge e. These terms precisely correspond to $\omega(d_{\mathrm{contr}}(\Gamma))$. $\square$

Remark 3.92 In [Idr19], we had claimed that c_{Γ_W} was always a number rather than a form on M. We thank Victor Turchin for pointing this out to us. The argument given in [Idr19] still works, as the degree counting argument in the last item of the previous proofs shows that $\deg c_{\Gamma_W}$ is greater than nv, rather than simply positive.

3.4.5 Partition Function as a Maurer–Cartan Element

As we already explained, the map $\omega : \mathsf{Graphs}'_R(U) \to \Omega^*_{\mathrm{PA}}(\mathsf{FM}_M(U))$ can unfortunately not be a quasi-isomorphism: the complex $\mathsf{Graphs}'_R(U)$ is too big. In the extreme case where $U = \varnothing$, we have $\mathsf{Conf}_M(\varnothing) = *$ so $\Omega^*_{\mathrm{PA}}(\mathsf{FM}_M(\varnothing)) = \mathbb{R}$ concentrated in degree zero. The CDGA $\mathsf{Graphs}'_R(\varnothing)$ is however far from being acyclic. It contains all the graphs composed only of internal vertices, and the cohomology of this complex is a priori non-trivial. If for example $R = \mathbb{R}$ (i.e. the graphs are not decorated), then Willwacher [Wil14] has for example shown that the CDGA contains in degree 0 a copy of the Grothendieck–Teichmüller Lie algebra $\mathfrak{grt}$ which is of infinite dimension. A copy of this algebra is included in all $\mathsf{Graphs}'_R(U)$.

We must therefore get rid of all *internal components* (i.e., composed exclusively of internal vertices). It is not possible to simply quotient by graphs that have internal components, as this would not be compatible with the integration procedure. In this section, we explain what the proper way of removing internal components is.

Definition 3.93 Let $\mathsf{fGC}_R = \mathsf{Graphs}'_R(\varnothing)$ be the CDGA of graphs without external vertices.

The product on fGC_R is simply the disjoint union of graphs. The CDGA fGC_R is thus quasi-free, i.e., it is freely generated as an algebra, by the sub-module of connected graphs GC_R. It moreover acts on every $\mathsf{Graphs}'_R(U)$ by disjoint union.

The differential $d : \mathsf{fGC}_R \to \mathsf{fGC}_R$ is quadratic-linear in terms of generators GC_R, i.e., for any $\gamma \in \mathsf{GC}_R$, the differential

$$d\gamma = d_1\gamma + d_2\gamma \in \mathsf{GC}_R \oplus S^2(\mathsf{GC}_R). \tag{3.71}$$

Indeed, the internal differential d_R and the contracting differential d_{contr} preserve the connectedness (i.e., they are linear), while the cutting differential transforms a connected graph into a linear combination of connected graphs (linear) and graphs with two connected components (quadratic).

Let us then consider the dual module $\mathsf{GC}_R^\vee$ and its suspension $\mathsf{GC}_R^\vee[-1]$. Koszul duality between commutative algebras and Lie algebras leads then to the following fact:

Proposition 3.94 *The dg-module* $\mathsf{GC}_R^\vee[-1]$ *is a differential-graded Lie algebra. The differential of* $\mathsf{GC}_R^\vee[-1]$ *is dual to the linear part of the differential* d, *while the Lie bracket of is dual to the quadratic part of* d.

More precisely, for $x, y \in \mathsf{GC}_R^\vee[-1]$ and $\gamma \in \mathsf{GC}_R$, we define the elements of the dual $\delta x \in \mathsf{GC}_R^\vee[-1]$ and $[x, y] \in \mathsf{GC}_R^\vee[-1]$ by:

$$\langle \delta x, \gamma \rangle := \langle x, d_1\gamma \rangle, \qquad \langle [x, y], \gamma \rangle := \langle x \otimes y, d_2\gamma \rangle. \tag{3.72}$$

Proof This classical result follows from the fact that $d : \mathsf{fGC}_R \to \mathsf{fGC}_R$ is a square-zero derivation implies that δ and $[-, -]$ satisfy the relations that define a dg-Lie algebra. Note that since $d^2 = (d_1 + d_2)^2 = 0$, by inspecting in which factors the linear maps land, we see that $d_1^2 = 0$, $d_1d_2 + d_2d_1 = 0$, and $d_2^2 = 0$.

- The relation $\delta^2 = 0$ simply follows from $d_1^2 = 0$.
- Graded antisymmetry of the bracket stems from the fact that d_2 lands in the symmetric product of GC_R.

- The Leibniz identity, $\delta[x, y] = [\delta x, y] + (-1)^{|x|}[x, \delta y]$, follows from $d_1 d_2 + d_2 d_1 = 0$. The map $x \otimes y \mapsto \delta[x, y]$ is dual to $d_2 d_1$, while $x \otimes y \mapsto [\delta x, y] + (-1)^{|x|}[x, \delta y]$ is dual to $d_1 d_2$.
- Finally, the Jacobi identity is dual to the identity $d_2^2 = 0$.

$\square$

We can interpret this dg-Lie algebra structure graphically. Elements of $\mathsf{GC}_R^\vee[-1]$ can be represented as potentially infinite sums of connected graphs decorated by $R^\vee$. These sums be seen as elements of the dual base of GC_R given by the graphs decorated by R. The differential $\delta : \mathsf{GC}_R^\vee[-1] \to \mathsf{GC}_R^\vee[-1]$ is the sum of several terms:

- the dual of the internal differential $\delta_{R^\vee}$, which acts on the decorations by a derivation;
- the "de-contracting" differential $\delta_{\mathrm{decontr}}$: it is the sum on all the ways to transform an internal vertex into an edge and to reconnect the edges incident to one or the other vertices; on the decorations, this differential acts by the co-product $R^\vee \to R^\vee \otimes R^\vee$ dual to the product $R \to R$;
- the "connecting" differential δ_{connec}: is the sum on all ways to connect any two vertices by an edge, sending the decorations α, β from the ends of the edges to the element of $R^\vee \otimes R^\vee$ defined by $x \otimes y \mapsto \langle \alpha \otimes \beta, (x \otimes y)\Delta_R \rangle$.

Finally, the bracket $[\gamma, \gamma']$ of two graphs γ, γ' is defined in a similar way to the connecting differential. It is the sum over all the ways to connect a vertex of γ with a vertex of γ', acting on the decorations as above.

Definition 3.95 The set of *Maurer–Cartan elements* $\mathrm{MC}(\mathfrak{g})$ of a Lie algebra $\mathfrak{g}$ is defined as the set of elements $x \in \mathfrak{g}^1$ satisfying $dx + \frac{1}{2}[x, x] = 0$.

The general theory then leads to the following:

Proposition 3.96 *The data of a morphism of CDGAs $f : \mathrm{fGC}_R \to \mathbb{R}$ is equivalent to the data of a Maurer–Cartan element $z \in \mathrm{MC}(\mathsf{GC}_R^\vee[-1])$.*

Proof This element z is simply the restriction of f to connected graphs. The compatibility of f with the differential and the product is encoded in the Maurer–Cartan relation. $\square$

Definition 3.97 The *partition function* of M is the morphism $z : \mathrm{fGC}_R \to \mathbb{R}$ equal to $\omega : \mathsf{Graphs}_R'(\varnothing) \to \Omega_{\mathrm{PA}}^*(\mathrm{FM}_M(\varnothing)) = \mathbb{R}$.

The terminology "partition function" comes from mathematical physics. This partition function is related to Chern–Simons invariants [AS94; BC98; CM10]. By Proposition 3.96, the data of the morphism $z : \mathrm{fGC}_R \to \mathbb{R}$ is equivalent to the data of a Maurer–Cartan element, still denoted $z \in \mathrm{MC}(\mathsf{GC}_R^\vee)$ by abuse of notation.

Recall that fGC_R acts on $\mathsf{Graphs}_R(U)$ by disjoint union. It also acts on $\mathbb{R}$ through the morphism z. We can thus make the following definition:

Definition 3.98 The reduced graph complex $\mathsf{Graphs}_R(U)$ is the tensor product $\mathsf{Graphs}_R'(U) \otimes_{\mathrm{fGC}_R} \mathbb{R}$.

Concretely, $\mathsf{Graphs}_R(U)$ is the quotient of $\mathsf{Graphs}'_R(U)$ by the ideal generated by the following relations: if $\Gamma \in \mathsf{Graphs}'_R(U)$ and $\gamma \in \mathrm{fGC}_R$, then $\Gamma \sqcup \gamma$ is identified with $z(\gamma) \cdot \Gamma$.

Proposition 3.99 *The map ω induces a CDGA morphism:*

$$\omega : \mathsf{Graphs}_R(U) \to \Omega^*_{\mathrm{PA}}(\mathsf{FM}_M(U)). \tag{3.73}$$

Proof Let $\Gamma \in \mathsf{Graphs}'_R(U)$ and $\gamma \in \mathrm{fGC}_R$ be two graphs. We must check that $\omega(\Gamma \sqcup \gamma) = z(\gamma)\omega(\Gamma)$. We just have to check that $\omega(\varnothing \sqcup \gamma) = z(\gamma)$, because $\Gamma \sqcup \gamma = \Gamma \cdot (\varnothing \sqcup \gamma)$ and ω is compatible with the product. This is essentially derived from Fubini's theorem: the map from $\mathsf{FM}_M(I)$ to the fiber of the projection $\mathsf{FM}_M(U \sqcup I) \to \mathsf{FM}_M(U)$ and is of degree 1. $\qquad\square$

Remark 3.100 Note that $\mathsf{Graphs}_R(U)$ is good from a homological point of view. Indeed, it is free as a graded commutative algebra. Generators are the internally connected graphs, i.e., the graphs that remain connected when all external vertices (but not their incident edges) are deleted.

3.4.6 Simplification of the Partition Function

We now want to define a morphism $\mathsf{Graphs}_R(U) \to \mathsf{G}_A(U)$ in order to complete the zigzag between $\mathsf{G}_A(U)$ and $\Omega^*_{\mathrm{PA}}(\mathsf{FM}_M(U))$. The natural idea would be to simply quotient by the graphs having internal vertices. Unfortunately, this operation is not compatible with the differential. Indeed, the cutting part can disconnect an internal component, which is then evaluated by the partition function to give a number. This number is usually not zero.

However, a part of the partition function is compatible with the differential. On a γ graph with a single vertex indexed by $x \in R$, the partition function is $z(\gamma) = \int_M x = \varepsilon(x)$. Combined with Lemma 3.12 and Eq. (3.64), we see that this identification is compatible with the quotient. Only terms where graphs with more than two vertices appear are problematic. This leads to the following definition:

Definition 3.101 The *trivial partition function* is the map $z_0 : \mathrm{fGC}_R \to \mathbb{R}$ given on generators by:

$$z_0(\gamma) := \begin{cases} \int_M x = \varepsilon(x), & \text{if } \gamma = \overset{x}{\bullet}; \\ 0, & \text{otherwise.} \end{cases} \tag{3.74}$$

We will show that the partition function z is homotopic to z_0 in the following sense.

Definition 3.102 Let A be a CDGA. Its *path CDGA* is the CDGA $A \otimes S(t, dt)$ obtained by adjoining two variables t (of degree 0) and dt (of degree 1) with $d(t) = dt$ and $d(dt) = 0$. There are two canonical morphisms $\mathrm{ev}_0, \mathrm{ev}_1 : A \otimes S(t, dt) \to A$ defined by:

$$\mathrm{ev}_i\big(a \otimes P(t)\big) = P(i) \cdot a \qquad\qquad \mathrm{ev}_i\big(a \otimes P(t)dt\big) = 0 \qquad (3.75)$$

Definition 3.103 Two morphisms $f, g : A \to B$ of CDGAs are *homotopic* if there exists a morphism of CDGAs $H : A \to B \otimes S(t, dt)$ such that $\mathrm{ev}_0 \circ H = f$ and $\mathrm{ev}_1 \circ H = g$.

Remark 3.104 Concretely, a homotopy $H : A \to B \otimes S(t, dt)$ is of the form:

$$H(a) = H_0(a) + H_1(a)t + H_2(a)t^2 + \cdots + H_0'(a)dt + H_1'(a)tdt + H_2'(a)t^2dt + \ldots \qquad (3.76)$$

where $H_0(a) = f(a)$, $\sum_{i \geq 0} H_i(a) = g(a)$, and H satisfies a number of relations such that $H(ab) = H(a)H(b)$ and $H \circ d = d \circ H$ (for example $H_1(ab) = f(a)H_1(b) + H_1(a)f(b)$, $H_0'(da) = f(a)$, etc).

Remark 3.105 Morphisms $\mathrm{fGC}_R \to \mathbb{R}$ are in bijection with Maurer–Cartan elements in the Lie algebra $\mathrm{GC}_R^\vee[-1]$ thanks to Proposition 3.94. The notion of homotopy of morphisms $\mathrm{fGC}_R \to \mathbb{R}$ is equivalent to the notion of gauge equivalence of Maurer–Cartan elements. If we unpack the definition, we see that two Maurer–Cartan elements $z_0, z_1 \in \mathrm{GC}_R^\vee[-1]$ are gauge equivalent if there exists an element ξ such that:

$$z_1 = e^{\mathrm{ad}\,\xi} z_0 - \frac{e^{\mathrm{ad}\,\xi} - \mathrm{id}}{\mathrm{ad}\,\xi}(d\xi), \qquad (3.77)$$

where $\mathrm{ad}\,\xi = [\xi, -]$ is the adjoint action of $\mathrm{GC}_R^\vee[-1]$ on itself.

In order to show that z is homotopic to z_0, we will need the following property of the propagator.

Proposition 3.106 ([CM10, Lemma 3]) . *The propagator φ can be chosen so that the following equation is true for any $x \in R$:*

$$\omega\left(\underset{1}{\bigcirc}\overset{x}{\rule{1.5em}{0.4pt}}\bullet\right) = 0. \qquad (3.78)$$

Proof We simply replace φ by

$$\varphi - \int_3 p_{13}^*(\varphi)p_{23}^*(\Delta_R) - \int_3 p_{23}^*(\varphi)p_{13}^*(\Delta_R) + \int_{3,4} p_{34}^*(\varphi)p_{13}^*(\Delta_R)p_{14}^*(\Delta_R), \qquad (3.79)$$

where the notation $\int_{i,j,k\ldots}$ represents the integral along the fibers of the application $FM_M(n) \to FM_M(r)$ that forgets the points $i, j, k \ldots$. $\qquad\square$

From now on, we assume that the propagator has been chosen so that this equation holds. This will have the following effect on the partition function:

Corollary 3.107 *The partition function z vanishes on all graphs containing a univalent vertex.*

Proof Let γ be such a graph, and let i be its univalent vertex, $x \in R$ its decoration, and j the only vertex adjacent to i. The partition function $z(\gamma)$ is calculated as the integral $\int_{FM_M(I)} \omega'(\gamma)$. The form $\omega'(\gamma)$ splits into $\omega'(\gamma') \wedge p_{ij}^*(\varphi) \wedge p_i^*(x)$. The integral on $FM_M(I)$ can be computed (thanks to a classical formula) by first computing the integral along the fibers of the bundle $FM_M(I) \to FM_M(I \setminus \{i\})$ then by computing the integral on $FM_M(I \setminus \{i\})$. The integral along the fibers in question is exactly the one that appears in Proposition 3.106 and is therefore zero. $\qquad\square$

The article [CM10] contains an additional property: they claim that the propagator φ can be chosen so that the partition function vanishes on any graph having a bivalent vertex decorated by $1 \in R$. This would allow to conclude that the partition function is equal to the trivial function for $n \geq 4$.

Proposition 3.108 *Let $\gamma \in GC_R$ be a graph having at least two vertices and containing no bivalent vertices decorated by the unit $1 \in R^0$. Suppose that $n \geq 4$. Then $z(\gamma) = 0$.*

Proof If γ contains a univalent vertex then we conclude with the previous corollary. Let us thus assume that γ contains no such vertex. Let $i + j$ be the number of vertices of γ, where j is the number of bivalent vertices (necessarily decorated by an element of degree ≥ 2, because R is 1-connected) and i the number of vertices at least trivalent. The number of edges of γ is at least equal to $\frac{3i+2j}{2}$. These edges are all of degree $n - 1$. Moreover, the decorations of the bivalent vertices contribute at least 2 to the degree each. It follows that the form $\omega'(\gamma)$ is of degree at least $\frac{3i+2j}{2}(n - 1) + 2j$. This form is integrated on the space $FM_M(i + j)$, which is of dimension $(i + j)n$. The difference between these two numbers satisfies:

$$\deg \omega'(\gamma) - \dim FM_M(i + j) \geq \frac{3i + 2j}{2}(n - 1) + 2j - (i + j)n$$

$$= \frac{1}{2}\big((i + j)(n - 3) - j(n - 5)\big). \tag{3.80}$$

This last expression is strictly positive if $n \geq 4$ and $i + j \geq 2$. The form $\omega'(\gamma)$ is thus of degree higher than the dimension of the space on which it is integrated, it is thus necessarily zero. $\qquad\square$

Unfortunately, the result of [CM10] does not apply to the SA framework. They use an operation $d_M : \Omega_{PA}^*(M \times N) \to \Omega_{PA}^*(M \times N)$ which only differentiates the

M coordinates (with $d = d_M + d_N$ and d_N). This operation is easily defined for de Rham forms, as it is enough to express it in local coordinates. But it would seem that it does not exist for PA forms. More precisely, it would be necessary to find an operation $d_M : \Omega^*_{PA}(M \times N) \to \Omega^{*+1}_{PA}(M \times N)$ such that if $\pi : M \times N \to M$ is the projection, then $d(\pi_*\alpha) = \pi_*(d_M\alpha)$. This does not seem possible based on some examples, e.g., $(x^2 + y^2)^{1/2} \in \Omega^0_{PA}([0, 1])$.

However, we can get around this difficulty in the following manner. The rough idea is that graphs containing bivalent vertices indexed by 1_R do not contribute to the cohomology of GC_R.

Definition 3.109 Let fGC^0_R be the quotient of fGC_R by the relation $\gamma \sim z(\gamma)$ for γ a graph with a single vertex.

Note that $z \in GC^\vee_R$ obviously factors through GC^0_R, since $z : fGC_R \to \mathbb{R}$ is a morphism of CDGAs. This factorization produces an element $\bar{z} \in (GC^0_R)^\vee$.

Definition 3.110 Let $I \subset fGC^0_R$ be the dg-ideal generated by graphs having at least one bivalent vertex decorated by $1 \in R$, an $fLoop_R \subset I$ be the dg-ideal generated by circular graphs, i.e., connected graphs without decorations whose vertices are all bivalent.

Proposition 3.111 *The inclusion* $fLoop_R \subset I$ *is a quasi-isomorphism.*

Proof (Sketch of Proof) This follows from a classical argument in the theory of graph complexes, see [Wil14, Proposition 3.4]. The idea is that given a graph $\gamma \in I$, one can defines its "core" as the (≥ 3)-valent graph obtained by forgetting all undecorated bivalent vertices and all antennas, i.e., sequences of undecorated bivalent vertices that terminates with a univalent vertex. See Fig. 3.6 for an example of a graph and its core.

One can then set up a spectral sequence (essentially by filtering with respect to the number of undecorated bivalent vertices) to show that only the circular graphs contribute to the cohomology of I. We refer to [Wil14, Proposition 3.4] for details.   □

Proposition 3.112 *The CDGA morphisms associated with z and z_0 are homotopic.*

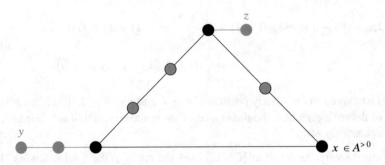

Fig. 3.6 A graph and its core (in black)

Proof The partition function cancels on circular graphs for degree reasons. By the previous proposition, $\bar{z}$ is thus homotopic to a morphism that vanishes on I, which can equivalently be interpreted as the fact that $\bar{z}$ is gauge equivalent to an element in $(\mathsf{GC}_R^0/I)^\vee$. Thanks to the argument of Proposition 3.108, this dg-Lie algebra vanishes in degree zero. It follows that $\bar{z}$ is homotopic to zero, i.e., z is homotopic to z_0. □

We defined $\mathsf{Graphs}_R(U)$ as the tensor product $\mathsf{Graphs}_R'(U) \otimes_{\mathsf{fGC}_R} \mathbb{R}$. We also know that the CDGA fGC_R is quasi-free and that the fGC_R-module $\mathsf{Graphs}_R'(U)$ is quasi-free. So if two morphisms $\mathsf{fGC}_R \to \mathbb{R}$ are homotopic, then the associated CDGAs $\mathsf{Graphs}_R'(U) \otimes_{\mathsf{fGC}_R} \mathbb{R}$ are quasi-isomorphic. This allows us to deduce the following from general arguments:

Proposition 3.113 *The CDGA* $\mathsf{Graphs}_R(U)$ *is quasi-isomorphic to* $\mathsf{Graphs}_R^0(U)$, *the CDGA which is constructed in the same way as* $\mathsf{Graphs}_R(U)$ *but using* z_0 *instead of* z *in the tensor product.*

3.5 End of the Proof

We can now tackle the last part of the proof. We first show easily that the quotient map $\mathsf{Graphs}_R^0(U) \to \mathsf{G}_A(U)$ is compatible with the differential and the product. We will then show that it is a quasi-isomorphism. Thanks to the theorem of Lambrechts and Stanley [LS08a], we know that the Betti numbers of $\mathsf{G}_A(U)$ are the same as those of $\mathsf{FM}_M(U)$. It will then be enough to show that $\mathsf{Graphs}_R(U) \to \Omega_{\mathrm{PA}}^*(\mathsf{FM}_M(U))$ is surjective in cohomology.

Remark 3.114 In this section, we use spectral sequences extensively. We refer to e.g., McCleary [McC01] for a survey. In short, we are mainly going to use the following approach. Let C and D be filtered complexes, i.e., $C = \bigcup_s F_s C$ where $\cdots \subset F_s C \subset F_{s+1} C \subset \ldots$ is a sequence of sub-complexes that starts at some $s_0 \in \mathbb{Z}$ and which is bounded above in each degree, and similarly for D. The E^0 page of the spectral sequence associated to C is $E^0 C = \bigoplus_s F_s C / F_{s-1} C$ equipped with the differential induced by the differential of C, and similarly for D. Let $f : C \to D$ be a cochain map which preserves the filtration, so that it induces a cochain map $E^0 f : E^0 C \to E^0 D$. If $E^0 f$ is a quasi-isomorphism, then so is f.

Definition 3.115 Let $\mathsf{Graphs}_A^0(U)$ be the quotient of $\mathsf{Graphs}_R^0(U)$ obtained by replacing all the decorations by elements of A.

Lemma 3.116 *There is a morphism of CDGAs* $\pi : \mathsf{Graphs}_A^0(U) \to \mathsf{G}_A(U)$ *given by modding out all internal vertices (where we use the graphical interpretation of* $\mathsf{G}_A(U)$ *written after Definition 3.15).*

Proof It is not hard to see that this is a morphism of algebras. We can also see that π vanishes on the image of the contracting differential. It can remove at most one internal vertex, so we need only to check this on graphs with exactly one internal vertex. There are three cases:

- If that internal vertex is univalent, then the contracting part of the differential cancels out with the cutting part of the differential, thanks to Eq. (3.64) and Lemma 3.12.
- If that internal vertex is bivalent, the contracting part of the differential appears twice, and the two terms cancel out thanks to the symmetry relations that define $G_A(U)$.
- Finally, if that internal vertex is at least trivalent, then π vanishes on the image of the differential thanks to a combination of the Arnold relations and the symmetry relations.

Finally, we must check that π is compatible with the cutting part of the differential. For edges between external vertices, this is clear by definition. For edges incident to internal vertices, this is also clear, except if cutting that edge completely separates an internal vertex from the rest of the graph. But since we now use z_0 instead of z to define $\mathsf{Graphs}_A^0(U)$, this is zero except if that internal component consists of a single vertex, in which case we saw above that this cancels out with part of the contracting part of the differential.   □

Lemma 3.117 *The map* $\mathsf{Graphs}_R^0(U) \to \mathsf{Graphs}_A^0(U)$ *is a quasi-isomorphism.*

Proof The two complexes are filtered by the number of edges. On the page E^0 of the associated spectral sequence, only the internal differentials of R and A remain. The two pages E^1 are thus given by $\mathsf{Graphs}_{H^*(M)}^0(U)$ and the map between the two is the identity. By classical theorems on spectral sequences (see McCleary [McC01]), the initial morphism is thus a quasi-isomorphism.   □

The following proposition will occupy us almost until the end of the section:

Proposition 3.118 *The quotient map* $\pi : \mathsf{Graphs}_A^0(U) \to G_A(U)$ *is a quasi-isomorphism.*

Let us filter the two complexes. We consider the filtration by the number of edges minus the number of vertices, that is to say, for some integer s:

$$F_s\mathsf{Graphs}_A^0(U) := \mathbb{R}\langle \Gamma \mid \#E_\Gamma - \#V_\Gamma \le s \rangle. \tag{3.81}$$

We can consider a similar filtration $F_s G_A(U)$ of $G_A(U)$, with the graphical interpretation of $G_A(U)$ written after Definition 3.15. More concretely, a monomial $\omega_{i_1 j_1} \ldots \omega_{i_k j_k} \in G_A(U)$ is said to live in filtration level $s = k - \#U$. The relations of $G_A(U)$ preserve filtration, i.e., they are homogeneous. Moreover, it is clear that the quotient map preserves the filtration.

We consider the spectral sequence associated to these filtered complexes. The complex $E^0 G_A(U)$ is the quotient complex $\bigoplus_s F_s G_A(U)$. Only the part of the differential which preserves the filtration level is thus kept. The cutting part of the differential $(d\omega_{ij} = \Delta_{ij})$ strictly decreases the filtration, so we are only left with the internal differential of A.

Similarly, on the complex $E^0\mathsf{Graphs}^0_A(U)$, we need to find out which parts of the differential preserve the filtration. Upon inspection, there only remains the internal differential, the contracting differential, and the cutting part of the differential which disconnects a univalent internal vertex. This cutting part cancels with the part of the differential that contracts the edge in question thanks to Lemma 3.12.

We can therefore split the two complexes $E^0\mathsf{G}_A(U)$ and $E^0\mathsf{Graphs}^0_A(U)$ in terms of the partition of U by connected components. Indeed, the remaining parts of the differential preserve connectedness.

Definition 3.119 Let $E^0\mathsf{Graphs}^0_A\langle U\rangle$ (resp. $E^0\mathsf{G}_A\langle U\rangle$) be the sub-complex constituted by connected graphs.

Lemma 3.120 *There is a splitting, where the sum runs over all partitions π of U:*

$$E^0\mathsf{Graphs}^0_A(U) \cong \bigoplus_\pi \bigotimes_{V\in\pi} E^0\mathsf{Graphs}^0_A\langle V\rangle. \tag{3.82}$$

A similar splitting exists for $E^0\mathsf{G}_A(U)$. The quotient map $E^0\mathsf{Graphs}^0_A(U) \to E^0\mathsf{G}_A(U)$ preserves the splitting.

Using all this, we just have to show that:

Lemma 3.121 *The map $E^0\mathsf{Graphs}^0_A\langle U\rangle \to E^0\mathsf{G}_A\langle U\rangle$ is a quasi-isomorphism.*

Recall that we have a graphical interpretation of $H^*(\mathsf{Conf}_{\mathbb{R}^n}(U))$, see Theorem 2.88. Thanks to the symmetry relation, the CDGA $E^0\mathsf{G}_A\langle U\rangle$ is isomorphic to the tensor product of A with the connected part of $H^*(\mathsf{Conf}_{\mathbb{R}^n}(U))$. The notation in the following proposition will be clear in Chap. 5.

Proposition 3.122 *Let $\mathsf{e}_n^\vee\langle U\rangle$ be the submodule of $H^*(\mathsf{Conf}_{\mathbb{R}^n}(U))$ spanned by connected graphs. Then we have the following:*

$$\dim \mathsf{e}_n^\vee\langle U\rangle^i = \begin{cases} (\#U - 1)!, & \text{if } i = (n-1)(\#U-1); \\ 0, & \text{otherwise.} \end{cases} \tag{3.83}$$

Proof This is a classical result. A simple connected graph on $\#U$ vertices must have at least $\#U - 1$ edges. If such a graph has at least $\#U$ edges, then there is a cycle, and using the Arnold relations repeatedly we arrive at a double edge, which vanishes due to the relations in Theorem 2.88. The graded module $\mathsf{e}_n^\vee\langle U\rangle$ is thus only nonzero in degree $(n-1)(\#U-1)$. In that degree, it is easily seen that a basis is given by the words $\omega_{1i_1}\ldots\omega_{1i_{n-1}}$, where $(i_1,\ldots,i_{n-1})$ is a permutation of $\{2,\ldots,n\}$. $\square$

The morphism $E^0\mathsf{Graphs}^0_A\langle U\rangle \to E^0\mathsf{G}_A\langle U\rangle$ is clearly surjective in cohomology, as we can represent $a \otimes [\Gamma]$ by $p_1^*(a)\Gamma$ for any closed representative of Γ. So we just have to show that Betti numbers coincide. We can compute these numbers by recurrence on $\#U$:

- if U is a singleton then $\dim H^i(E^0G_A\langle U\rangle) = \dim H^i(A)$;
- if $\#U \geq 2$ and $u \in U$, then

$$\dim H^i(E^0G_A\langle U\rangle) = (\#U - 1) \cdot \dim H^{i-n+1}(E^0G_A\langle U \setminus \{u\}\rangle). \qquad (3.84)$$

Lemma 3.123 *The Betti numbers of $E^0\mathsf{Graphs}^0_A\langle U\rangle$ satisfies the same relations as Eq. (3.83).*

Proof Let us start with the case $\#U = 1$. There is an explicit homotopy which shows that the complex $E^0\mathsf{Graphs}^0_A\langle U\rangle$ has the same cohomology as A. It can be represented graphically by:

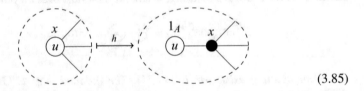

$$(3.85)$$

Then a small calculation shows that $(dh + hd)(\Gamma) = \Gamma$ unless Γ has no internal vertices, in which case $(dh + hd)(\Gamma) = 0$. This shows that cohomology classes of $E^0\mathsf{Graphs}^0_A\langle U\rangle$ are exactly the cohomology classes of graphs with no internal vertices, which form a complex isomorphic to A.

Now, let U be a finite set with at least two elements and let us fix $u \in U$. Let us denote by $C_u \subset E^0\mathsf{Graphs}^0_A\langle U\rangle$ the sub-complex generated by graphs such that u is univalent, decorated by $1 \in A$ and connected to another external vertex. Clearly,

$$C_u \cong \bigotimes_{v\in U,\ v\neq u} e_{uv} \cdot E^0\mathsf{Graphs}^0_A\langle U \setminus \{u\}\rangle. \qquad (3.86)$$

This subcomplex satisfies the correct relations on Betti numbers. Moreover, the inclusion $C_u \subset E^0\mathsf{Graphs}^0_A\langle U\rangle$ is a quasi-isomorphism. Indeed, if we let Q be the quotient $E^0\mathsf{Graphs}^0_A\langle U\rangle/C_u$, we have $Q = Q_1 \oplus Q_2$ where

- Q_1 is the sub-module generated by graphs where u is univalent, decorated by 1, and connected to an internal vertex,
- Q_2 is generated by the other kinds of graphs, i.e., u is either at least bivalent or decorated by some $a \in A^{\geq 2}$.

We can filter Q by $F_sQ_1 = \{\Gamma \mid \#E_\Gamma \leq s + 1\}$ and $F_sQ_2 = \{\Gamma \mid \#E_\Gamma \leq s\}$. Then the differential on the page E^0 of the associated spectral sequence maps E^0Q_1 to E^0Q_2 by an isomorphism. This shows that Q is acyclic. It follows that $C_u \simeq E^0\mathsf{Graphs}^0_A\langle U\rangle$, which concludes the proof. $\square$

Proof (End of the Proof of Proposition 3.118) The morphism $E^0\mathsf{Graphs}_A\langle U\rangle \to$ $E^0\mathsf{G}_A\langle U\rangle$ is clearly surjective in cohomology. It follows that the quotient map induces a quasi-isomorphism on the page E^0 and therefore that it is a quasi-isomorphism.   □

It only remains to show that:

Lemma 3.124 *The application* $\mathsf{Graphs}_A(U) \to \Omega^*_{\mathrm{PA}}(\mathsf{FM}_M(U))$ *is surjective in homology*

Proof This simply requires to represent any class by a graph, which is easily done.   □

We have almost finished proving the theorem. However, so far, we have set the model to Poincaré duality. We are thus left with proving the following proposition:

Proposition 3.125 *If A and B are two quasi-isomorphic Poincaré duality CDGAs, then* $\mathsf{G}_A(U)$ *and* $\mathsf{G}_B(U)$ *are as well.*

Proof The proof still involves graph complexes, but does not use transcendental methods (i.e., integrals). Let us fix a zigzag of quasi-isomorphisms $A \leftarrow R \to B$. We then have two maps $\varepsilon_A, \varepsilon_B \ : \ R^n \ \to \ \mathbb{R}$ using the augmentations of A and B. One thus obtains two graph complexes $\mathsf{Graphs}^{\varepsilon_A}_R(U)$ and $\mathsf{Graphs}^{\varepsilon_B}_R(U)$ defined as $\mathsf{Graphs}^0_R(U)$ was defined previously but using ε_A and ε_B instead. The quotient maps $\mathsf{Graphs}^{\varepsilon_A}_R(U) \ \to \ \mathsf{G}_A(U)$ and $\mathsf{Graphs}^{\varepsilon_B}_R(U) \ \to \ \mathsf{G}_B(U)$ are quasi-isomorphisms from the of Proposition 3.118 (which did not involve integrals). It is therefore simply enough to show that the two graph complexes are quasi-isomorphic. Up to multiplying one of the two augmentations by a scalar (which induces an automorphism of the associated graph complex), we can assume that ε_A and ε_B induce the same application in cohomology. The two maps are thus homotopic as chain maps $R \ \to \ \mathbb{R}[-n]$. That homotopy induces a homotopy between the CDGA morphisms $z_A, z_B \ : \ \mathsf{fGC}_R \ \to \ \mathbb{R}$. By reusing the techniques developed earlier, we deduce that $\mathsf{Graphs}^{\varepsilon_A}_R$ and $\mathsf{Graphs}^{\varepsilon_B}_R$ are quasi-isomorphic.   □

Combining everything we have just seen, we have finished demonstrating Theorem 3.20 for dim $M \geq 4$. The case dim $M < 4$ is treated completely differently. Note that two homotopy equivalent closed manifolds have the same dimension.

- In dimension 0, the only simply connected closed manifold is the singleton $\mathbb{R}^0 = \{0\}$. The result is then rather clear (or perhaps badly defined because $\mathsf{G}_A(U)$ contains elements of negative degree). Indeed, $\mathsf{Conf}_{\mathbb{R}^0}(U)$ is empty for $\#U \geq 2$. One chooses the CDGA $H^*(\mathbb{R}^0) = (\mathbb{R}, d = 0)$ as Poincaré duality model of $\mathbb{R}^0$, with $\Delta_{\mathbb{R}} = 1 \otimes 1$. The Lambrechts–Stanley model then becomes

$$G_{H^*(\mathbb{R}^0)}(U) = \big(S(\omega_{ij})_{i\neq j\in U}/I, d\omega_{ij} = 1\big).$$

This CDGA is clearly acyclic for $\#U \geq 2$. Indeed, if α is a cocycle, then $\alpha = d(\omega_{ij}\alpha)$ is automatically a coboundary (where $i \neq j \in U$).

- In dimension 1, there is no simply connected closed manifold.
- In dimension 2, the only simply connected closed manifold is the sphere $\mathbb{S}^2$. This sphere is also the complex projective line $\mathbb{CP}^1$, which is a smooth complex projective manifold. We can therefore apply the result of Kriz [Kri94], which says that for a smooth complex projective manifold M, $\mathsf{G}_{H^*(M)}(U)$ is indeed a rational (thus real) model of $\mathsf{Conf}_M(U)$. [[CW16], Appendix B] have also explicitly computed the partition function in this case and shown that it is trivial by analytical methods.
- In dimension 3, thanks to the Poincaré conjecture (now a theorem of Perelman [Per02; Per03], see also Morgan and Tian [MT07] and Kleiner and Lott [KL08]) the only simply connected closed manifold is the sphere $\mathbb{S}^3$. This sphere is a Lie group. Specifically, it is the unitary special group $SU(2)$. Recall that for a Lie group G, the Fadell–Neuwirth fibration:

$$\mathsf{Conf}_{G\backslash *}(r) \hookrightarrow \mathsf{Conf}_G(r+1) \twoheadrightarrow G \qquad (3.87)$$

which keeps only the first point splits thanks to a result of Fadell and Neuwirth [FN62a], see Remark 2.91.

As the sphere is formal, one can choose its cohomology $A = H^*(\mathbb{S}^3) = S(v)/(v^2)$ (with $\deg v = 3$) to take the role of Poincaré duality model. In our case, $\mathbb{S}^3 \setminus * = \mathbb{R}^3$, and we know that the configuration spaces of $\mathbb{R}^3$ are formal (Sect. 2.4). Combined with the previous splitting, it is thus enough to show that $\mathsf{G}_A(U \sqcup \{0\})$ is a model of $\mathbb{S}^3 \times \mathsf{Conf}_{\mathbb{R}^3}(U)$ for any U. There is an explicit quasi-isomorphism (we thank Thomas Willwacher for pointing this out):

$$\pi : A \otimes H^*(\mathsf{Conf}_{\mathbb{R}^3}(U)) \to \mathsf{G}_A(U \sqcup \{0\}),$$

$$v \otimes 1 \mapsto p_0^*(v),$$

$$1 \otimes \omega_{ij} \mapsto \omega_{ij} - \omega_{0i} - \omega_{0j}. \qquad (3.88)$$

Since the two CDGAs have the same cohomology, it is sufficient to verify that π is surjective in cohomology. As the cohomology of the CDGA at the target is generated in degrees 2 and 3, it is sufficient to verify that π is surjective in these degrees. It is clear in degree 3: indeed, if $p_0^*(v)$ were not a generator, we would have $p_0^*(v) = d\omega$ where ω is a sum of ω_{ij}, but all the $p_i^*(v)$ come in pairs in $d\omega_{ij}$. In degree 2, it is a matter of a little diagram chase, considering the quotient maps $\mathsf{G}_A(U \sqcup \{*\}) \to H^*(\mathsf{Conf}_{\mathbb{R}^3}(U))$ and $A \otimes H^*(\mathsf{Conf}_{\mathbb{R}^3}(U)) \to H^*(\mathsf{Conf}_{\mathbb{R}^3}(U))$.

Chapter 4
Configuration Spaces of Manifolds with Boundary

The results of the previous chapter apply to closed manifolds, that is, compact manifolds without boundary. In this chapter, we extend these to manifolds with boundary. The case of manifolds with boundary is more difficult than the case of manifolds: in general, the homotopy types of the configuration spaces of a manifold with boundary M depend on the homotopy type of the pair $(M, \partial M)$, not just the homotopy type of M.

We start this chapter with the motivation, namely, the computation of homotopy types of configuration spaces "by induction" (see Sect. 4.1). If a manifold is obtained by gluing two submanifolds along their boundaries, then its configuration spaces can be obtained from the configuration spaces of the submanifolds. Our objective is thus to be able to study configuration spaces of large manifolds through the study of smaller ones. We then present in Sect. 4.3 two different kinds of models (in the sense of real homotopy theory, see Sect. 2.3) of configuration spaces of manifolds with boundary. The starting data to define the models is Poincaré–Lefschetz duality, and we introduce the cochain version of that duality in Sect. 4.2. The first kind of models (Sect. 4.3) is one based on graph complexes, just like in Sect. 3.4. This graph-complex based model is appropriate for inductive computations, even though it is quite large. The second kind of models (Sect. 4.4) is inspired by the Lambrechts–Stanley model that we studied in the previous chapter. However, the algebraic relation between generators of the Lambrechts–Stanley model must be perturbed in order to get an actual CDGA model.

4.1 Motivation

A very common idea in algebraic topology is to cut a large manifold into simpler submanifolds, study those simpler manifolds separately, and glue back the information to recover information about the larger manifold. This idea can apply

© The Author(s), under exclusive license to Springer Nature Switzerland AG 2022
N. Idrissi, *Real Homotopy of Configuration Spaces*, Lecture Notes
in Mathematics 2303, https://doi.org/10.1007/978-3-031-04428-1_4

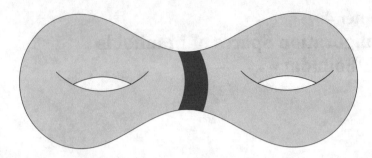

Fig. 4.1 The surface Σ_2 seen as $\Sigma_{1,1} \cup_{S^1 \times \mathbb{R}} \Sigma_{1,1}$

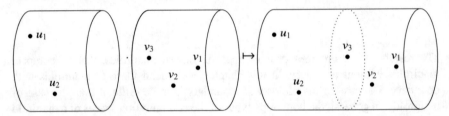

Fig. 4.2 The collection $\mathrm{Conf}_{N \times \mathbb{R}}$ is a monoid up to homotopy

to configuration spaces. To simplify, assume that a manifold X is given by the union of two submanifolds M and M' along their common boundary $\partial M = \partial M' = N$. There is then a formula that allows to express the configuration spaces of X in terms of those of M, M' and N, as we now explain. In order to deal with smoothness, let us also assume that we have chosen a tubular neighborhood $N \times \mathbb{R} \hookrightarrow X = M \cup_N M'$ such that $N \times \{0\}$ is the common boundary of M and M', $N \times \mathbb{R}_{>0}$ is included in M, and $N \times \mathbb{R}_{<0}$ is included in M' (see Fig. 4.1).

Under these hypotheses, the collection $\mathrm{Conf}_{N \times \mathbb{R}} = \{\mathrm{Conf}_{N \times \mathbb{R}}(U)\}_U$, where U ranges over all finite sets, is a monoid up to homotopy. To see this, let us choose an embedding $\mathbb{R} \sqcup \mathbb{R} \hookrightarrow \mathbb{R}$, for example by identifying the first factor with $\mathbb{R}_{<0}$ and the second with $\mathbb{R}_{>0}$. Then we have an induced embedding $N \times \mathbb{R} \sqcup N \times \mathbb{R} \hookrightarrow N \times \mathbb{R}$, which induces, for finite sets U and V, an embedding:

$$\mathrm{Conf}_{N \times \mathbb{R}}(U) \times \mathrm{Conf}_{N \times \mathbb{R}}(V) \hookrightarrow \mathrm{Conf}_{N \times \mathbb{R}}(U \sqcup V). \qquad (4.1)$$

This operation can be represented by Fig. 4.2. The collection of all such maps (as U and V range over all finite sets) satisfies compatibility relations with respect to bijections $(U, V) \cong (U', V')$. In addition, it satisfies an associativity relation up to homotopy, and a unit relation up to homotopy (the unit being the empty configuration).

Remark 4.1 The collection $\mathrm{Conf}_{N \times \mathbb{R}}$ does not form a strict monoid: associativity and unit are valid only up to homotopy (as in a loop space, for example). It is possible to use a point of view that solves this problem thanks to the theory of

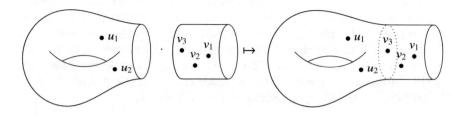

Fig. 4.3 The collection Conf_M is a right module over $\mathsf{Conf}_{N \times \mathbb{R}}$

operads (see Proposition 5.47). In Sect. 4.3, we define compactifications $\mathsf{aFM}_N(U)$ of $\mathsf{Conf}_{N \times \mathbb{R}}(U)$ such that $\mathsf{aFM}_N = \{\mathsf{aFM}_N(U)\}$ is a strictly associative and unital monoid.

Let us now describe how the monoid $\mathsf{Conf}_{N \times \mathbb{R}}$ interacts with the configuration spaces of M. The collection $\mathsf{Conf}_M = \{\mathsf{Conf}_M(U)\}$ is a right module (up to homotopy) over the monoid $\mathsf{Conf}_{N \times \mathbb{R}}$ (Fig. 4.3). To see this, recall the collar chosen above around $N = N \times \{0\} \subset N \times \mathbb{R}_{\geq 0} \subset M$. By using this collar, we obtain maps, illustrated in Fig. 4.3:

$$\mathsf{Conf}_M(U) \times \mathsf{Conf}_{N \times \mathbb{R}_{>0}}(V) \to \mathsf{Conf}_M(U \sqcup V). \tag{4.2}$$

The collection of all such maps endows Conf_M with the structure of a right module over $\mathsf{Conf}_{N \times \mathbb{R}_{>0}} \cong \mathsf{Conf}_{N \times \mathbb{R}}$.

Finally, in a similar manner, $\mathsf{Conf}_{M'}$ is a left module (up to homotopy) over the monoid $\mathsf{Conf}_{N \times R}$. The configuration spaces of $X = M \cup_N M'$ then satisfy the following formula, which uses the algebraic structures we have just defined:

$$\begin{aligned}
\mathsf{Conf}_X(U) &\simeq \left(\mathsf{Conf}_M \otimes_{\mathsf{Conf}_{N \times \mathbb{R}}} \mathsf{Conf}_{M'} \right)(U) \\
&:= \left(\bigsqcup_{U = V \sqcup V'} \mathsf{Conf}_M(V) \times \mathsf{Conf}_{M'}(V') \right) / \sim,
\end{aligned} \tag{4.3}$$

where the relation $\sim$ is given by $(x \cdot y, z) \sim (x, y \cdot z)$ for any $x \in \mathsf{Conf}_M$, $y \in \mathsf{Conf}_{N \times \mathbb{R}}$ and $z \in \mathsf{Conf}_{M'}$, just like in a classical tensor product.

In this chapter, we will define models for the real homotopy type of the objects $\mathsf{Conf}_{N \times \mathbb{R}}$ and Conf_M that take into account the above algebraic structures. The ultimate goal is to be able to compute the configuration spaces of large manifolds using Eq. (4.3).

Remark 4.2 The results below mostly concern simply connected manifolds and thus do not apply to closed surfaces other than $\mathbb{S}^2$. In [CIW19], with Campos and Willwacher, we have generalized the result of Sect. 4.4.2 to framed configuration spaces (cf. Sect. 5.3.2) of oriented surfaces.

Let us now give an outline of this chapter. In Sect. 4.2, we review Poincaré–Lefschetz duality models, a generalization of Poincaré duality models that apply to compact manifolds with boundary. In Sect. 4.3, we define models for the collection $\mathsf{Conf}_{N \times \mathbb{R}}$ and Conf_M as above, using the theory of graph complexes. In Sect. 4.4, we will simplify the model for Conf_M to obtain a "Lambrechts–Stanley"-like model. Unless otherwise stated, the results come from the article [CILW18], written in collaboration with Campos, Lambrechts, and Willwacher.

4.2   Poincaré–Lefschetz Duality Models

For closed manifolds, the construction of the Lambrechts–Stanley model of Chap. 3 was based on the notion of Poincaré duality CDGAs. These were defined through non-degenerate pairings $A^k \otimes A^{n-k} \to \mathbb{R}$ for $k \in \mathbb{Z}$. Every simply connected closed manifold admits such a model thanks to Theorem 3.6 [LS08b]. For compact manifolds with nonempty boundary, no such pairing can exist, as even the cohomology does not satisfy Poincaré duality. Instead, Poincaré duality is replaced by Poincaré–Lefschetz duality. Recall that if $(M, \partial M)$ is an oriented compact manifold with boundary of dimension n, then the evaluation on the fundamental class and the cup-product together induce non-degenerate pairings $H^k(M) \otimes H^{n-k}(M, \partial M) \to \mathbb{R}$, for all $k \in \mathbb{Z}$. In this section, we will explain how to adapt the definition of Poincaré duality CDGAs to this framework. We will start by some informal motivation and then introduce the precise definitions.

Let $(M, \partial M)$ be an oriented compact manifold with boundary of dimension n. Suppose that we are given a real model of the inclusion $i : \partial M \to M$, i.e., a CDGA morphism $\lambda : B \to B_\partial$ such that there is a zigzag of quasi-isomorphisms:

$$
\begin{array}{ccc}
B & \xleftarrow{\ \sim\ } \cdot \xrightarrow{\ \sim\ } & \Omega^*_{\mathrm{PA}}(M) \\
\Big\downarrow{\lambda} & \Big\downarrow & \Big\downarrow{i^*} \\
B_\partial & \xleftarrow{\ \sim\ } \cdot \xrightarrow{\ \sim\ } & \Omega^*_{\mathrm{PA}}(\partial M)
\end{array}
\tag{4.4}
$$

Then $\Omega^*_{\mathrm{PA}}(M, \partial M) := \ker(i^*)$ is quasi-isomorphic (as a mere cochain complex) to the homotopy kernel hoker λ, defined as follows.

Definition 4.3 Let $f : X \to Y$ be a cochain map. The *homotopy kernel* of f is the cochain complex:

$$
\mathrm{hoker}(f) := \left(X^* \oplus Y^{*-1}, \ d(x, y) = (dx, f(x) - dy) \right).
\tag{4.5}
$$

By general arguments, one can always find a model λ that is surjective. The following proposition is classical:

Proposition 4.4 *Let f be a cochain map. The kernel* ker f *embeds in* hoker f *through $x \mapsto (x, 0)$. If f is a surjection, then the inclusion i : ker $f \hookrightarrow$ hoker f is a quasi-isomorphism.*

Proof Let $x \in$ ker f be a cocycle such that $i(x)$ is a coboundary, i.e., there are elements $x' \in X$, $y' \in Y$ such that:

$$(x, 0) = d(x', y') = (dx', f(x') - dy'). \tag{4.6}$$

Then $f(x' - dy) = f(x') - dy' = 0$, so that $x' - dy'$ belongs to the kernel, and $x = dx' = d(x' - dy)$ is thus a coboundary in ker f. It follows that i is injective in cohomology.

Moreover, let $(x, y) \in$ hoker(f) be a cocycle, i.e., $dx = 0$ and $f(x) = dy$. Since f is surjective, there exists $x' \in X$ such that $f(x') = y$. Then

$$i(x - dx', 0) = (x - dx', 0) = (x, y) - d(x', 0). \tag{4.7}$$

Hence i is surjective in cohomology.   □

Moreover, we have the following classical result, which, together with the previous proposition, we can interpret as the fact that the homotopy kernel is a homotopy invariant version of the kernel.

Proposition 4.5 *Let $f : X \to Y$ be a cochain map. There is a functorial long exact sequence, where π :* hoker$(f) \to X$ *is the projection and ∂ : $Y[-1] \to$ hoker(f) is the inclusion:*

$$\cdots \to H^n(\text{hoker}(f)) \xrightarrow{\pi} H^n(X) \xrightarrow{f} H^n(Y) \xrightarrow{\partial} H^{n+1}(\text{hoker}(f)) \to \cdots \tag{4.8}$$

Corollary 4.6 *Suppose we are given a commutative square:*

$$
\begin{array}{ccc}
X & \xrightarrow[\alpha]{\sim} & X' \\
\downarrow{\scriptstyle f} & & \downarrow{\scriptstyle f'} \\
Y & \xrightarrow[\beta]{\sim} & Y'
\end{array}
\tag{4.9}
$$

where α and β are quasi-isomorphisms. Then the induced map hoker$(f) \to$ hoker(f') *is a quasi-isomorphism.*

Using Proposition 3.80, we have that $\Omega^*_{\text{PA}}(M) \to \Omega^*_{\text{PA}}(\partial M)$ is surjective. It thus follows from the above results that given Eq. 4.4, if we choose a surjective model λ, then $\Omega^*_{\text{PA}}(M, \partial M)$ is quasi-isomorphic to $K := \ker \lambda$. Poincaré–Lefschetz duality then tells us that the cohomology of B and that of K are dual to each other. If the boundary is non-empty, then it is not reasonable to look for a non-degenerate pairing between B and K. Indeed, as dim $K = $ dim $B - $ dim $B_\partial < $ dim B, it is impossible for K and B to be paired in a non-degenerate manner.

Let us find out what we could replace this pairing with. In a CDGA A with Poincaré duality, the pairing is induced by a form $\varepsilon : A^n \to \mathbb{R}$ which is compatible with the differential, i.e., $\varepsilon \circ d = 0$. Assuming that A is a model of a closed manifold W, the form ε represents the evaluation on the fundamental class $[W] \in H_n(W)$, or the integral on W. In a manifold with boundary $(M, \partial M)$, the integral of an exact form is not necessarily zero: by Stokes' formula, we know that $\int_M d\alpha = \int_{\partial M} \alpha|_{\partial M}$. Suppose that B and B_∂ are two CDGAs that are respectively models of M and ∂M, and that $\lambda : B \to B_\partial$ is a model of the inclusion. Then part of the Poincaré–Lefschetz duality structure of CDGA becomes the data of two forms, $\varepsilon : B^n \to \mathbb{R}$ and $\varepsilon_\partial : B_\partial^{n-1} \to \mathbb{R}$, which represent respectively the integrals on M and ∂M, and which must satisfy the relation $\varepsilon \circ d = \varepsilon_\partial \circ \lambda$.

The form ε_∂ induces pairings $B_\partial^k \otimes B_\partial^{n-1-k} \to \mathbb{R}$. We will ask that these pairings are non-degenerate for any $k \in \mathbb{Z}$. Moreover, the form ε induces pairings $B^i \otimes B^{n-i} \to \mathbb{R}$, but these are not compatible with the differential (i.e., $\langle da, b \rangle$ may not equal $\langle a, db \rangle$ up to sign). One can, however, restrict one of the two factors to K to obtain pairings $B^i \otimes K^{n-i} \to \mathbb{R}$ which are compatible with the differential. By dimension reasons, these pairings are necessarily degenerate for some i. We are then led to consider the quotient $P := B/I$ where I is the ideal of orphans, i.e.

$$I = \{b \in B \mid \langle b, k \rangle = 0, \; \forall k \in K\}. \tag{4.10}$$

The previous pairings go through the quotient, by definition, and induce new pairings $P^i \otimes K^{n-i} \to \mathbb{R}$. The next condition for a Poincaré–Lefschetz duality pair is that these new pairings are non-degenerate for all i, which mirrors the fact that $H^i(M) \otimes H^{n-i}(M, \partial M) \to \mathbb{R}$ is non-degenerate. The last condition requires that P remains a real model of M. We thus obtain the following definition:

Definition 4.7 A *Poincaré–Lefschetz duality pair* (PLD pair) is the data of a surjective CDGA morphism $B \xrightarrow{\lambda} B_\partial$ and two linear maps (called "orientations") $\varepsilon : B^n \to \mathbb{R}$, $\varepsilon_\partial : B_\partial^{n-1} \to \mathbb{R}$, satisfying the following properties:

- the pair $(B_\partial, \varepsilon_\partial)$ is a Poincaré duality CDGA (in particular $\varepsilon_\partial \circ d = 0$);
- Stokes' formula is satisfied, i.e., $\varepsilon \circ d = \varepsilon_\partial \circ \lambda$;
- let $K = \ker(\lambda)$, then the morphism $\theta_B : B \to K^\vee[-n]$ defined by $\theta_B(b)(k) := \varepsilon(bk)$ is a surjective quasi-isomorphism.

A *PLD model* of a manifold with boundary M is a PLD pair as above such that the morphism λ is a model of the inclusion $\partial M \hookrightarrow M$.

Remark 4.8 Poincaré–Lefschetz duality pairs are inspired by (and generalize) the surjective "pretty models" of Cordova Bulens et al. [CLS19].

The following proposition is obvious from the definition:

Proposition 4.9 *Let $(B \xrightarrow{\lambda} B_\partial, \varepsilon, \varepsilon_\partial)$ be a PLD pair. Let $P = B/I$ where $I = \ker \theta_B$. Then the quotient map $B \to P$ is a quasi-isomorphism. Moreover,* *the pairings $B^i \otimes K^{n-i} \to \mathbb{R}$ (for $i \in \mathbb{Z}$) given by $b \otimes k \mapsto \varepsilon(bk)$ go through the quotient and induces non-degenerate pairings $P^i \otimes K^{n-i} \to \mathbb{R}$.*

The objects involved in this proposition and the relationships between them can be summarized by the following diagram:

$$
\begin{array}{ccc}
K := \ker \lambda & \xleftarrow{\ \ \sim\ \ } & \Omega^*(M, \partial M) \\
\downarrow & & \downarrow \\
P := B/I \xleftarrow{\ \ \sim\ \ }_{\pi} B & \xleftarrow{\ \ \sim\ \ } & \Omega^*(M) \\
\downarrow{\scriptstyle \lambda} & & \downarrow{\scriptstyle i^*} \\
B_\partial & \xleftarrow{\ \ \sim\ \ } & \Omega^*(\partial M)
\end{array}
\tag{4.11}
$$

Example 4.10 Let $M = \mathbb{D}^n$ be the disk of dimension n. A surjective model of the inclusion $\partial M = \mathbb{S}^{n-1} \hookrightarrow \mathbb{D}^n$ is given by the pair $\lambda : B \to B_\partial$ where

- the CDGA $B_\partial = \mathbb{R}\langle 1, v \rangle$ is of dimension 2 with $\deg v = n - 1$ and $v^2 = 0$;
- the CDGA $B = \mathbb{R}\langle 1, v, w \rangle$ is of dimension 3 with $\deg w = n$ and $dv = w$ and all non-trivial products vanish;
- the morphism $\lambda : B \to B_\partial$ is defined by $\lambda(1) = 1$, $\lambda(v) = v$ and $\lambda(w) = 0$.

The orientations $\varepsilon : B^n \to \mathbb{R}$ and $\varepsilon_\partial : B_\partial^{n-1} \to \mathbb{R}$ are respectively given by $\varepsilon(w) = 1$ and $\varepsilon_\partial(v) = 1$. This defines a Poincaré–Lefschetz duality pair. The kernel $K = \ker \lambda$ is simply given by $\mathbb{R}\langle w \rangle$ concentrated in degree n. The map $\theta_B : B \to K^\vee[-n]$ is given by $\theta_B(1) = w^\vee$ and $\theta_B(v) = \theta_B(w) = 0$. We thus have $I = \mathbb{R}\langle v, w \rangle$ and $P = B/I = \mathbb{R}\langle 1 \rangle$. The pairing $K \otimes P \to \mathbb{R}$ simply pairs 1 and w.

Example 4.11 This example can be generalized to any manifold obtained by removing a disk from a closed manifold. If A is a Poincaré duality CDGA, then one can define a Poincaré–Lefschetz duality pair by $B_\partial = \mathbb{R}\langle 1, v_{n-1} \rangle$ and $B = A \oplus \mathbb{R}\langle v_{n-1} \rangle$ with $dv = \mathrm{vol}_A = \varepsilon^{-1}(1)$.

We now get to the main result of this section.

Proposition 4.12 ([CILW18, Proposition 2.5]) *Any simply connected compact manifold with simply connected boundary of dimension at least 7 admits a Poincaré–Lefschetz duality model.*

Proof The proof is inspired by that of the main theorem of [LS08b]. The idea is the following. Using standard arguments of rational homotopy theory, we can find a surjective morphism of CDGAs $\rho : R \to R_\partial$ which fits into a commutative diagram as follows:

$$
\begin{array}{ccc}
R & \xrightarrow{\ \ \sim\ \ } & \Omega^*(M) \\
\downarrow{\scriptstyle \rho} & & \downarrow{\scriptstyle \text{restriction}} \\
R_\partial & \xrightarrow{\ \ \sim\ \ } & \Omega^*(\partial M).
\end{array}
\tag{4.12}
$$

We can assume that R and R_∂ are simply connected ($R^0 = R_\partial^0 = \mathbb{R}$ and $R^1 = R_\partial^1 = 0$), that $R_\partial^2 \subset \ker d$, and that $(\ker \rho)^2 \subset \ker d$. Informally, we are going to think of R and R_∂ as cofibrant models of M and ∂M, and our goal will be to mod them out by some ideals to find a PLD model (B, B_∂).

The orientations are defined by:

$$\varepsilon : R \xrightarrow{\sim} \Omega^*(M) \xrightarrow{\int_M} \mathbb{R}, \tag{4.13}$$

$$\varepsilon_\partial : R_\partial \xrightarrow{\sim} \Omega^*(\partial M) \xrightarrow{\int_{\partial M}} \mathbb{R}. \tag{4.14}$$

These orientations satisfy the Stokes formula because the integrals on M and ∂M satisfy it, and they induce Poincaré–Lefschetz duality in cohomology. In general, though, they do not induce a duality at the level of cochains: there may exist elements $x \in R$ (resp. $y \in R_\partial$) such that $\varepsilon(x \cdot -) = 0$ (resp. $\varepsilon_\partial(y \cdot -) = 0$). These elements are called orphans. Thanks to [LS08b, Theorem 1.1], we can also replace R_∂ up to quasi-isomorphism by a Poincaré duality CDGA of dimension $n - 1$, so we may assume that there are no orphans in R_∂.

The next part of the proof is inspired by the proof of Theorem 3.6. The set of orphans forms an ideal. If we kill the orphans, then the quotient satisfies Poincaré–Lefschetz duality. However, the ideal of orphans is not generally acyclic, so the quotient does not necessarily have the correct homotopy type. Indeed, if o is an orphan cycle, we know (by Poincaré–Lefschetz duality in cohomology) that it is the boundary of a certain z. But that z may not itself be an orphan. The idea of [LS08b] is to add a new formal variable for each orphan, degree by degree, in a way that does not change the homotopy type but that makes the ideal of orphans acyclic in the extended algebra.

Let us proceed with the same idea. Let $K_R = \ker \rho$ and $\theta_R : R \to K^\vee[-n]$ be the morphism given by $\theta_R(x)(y) = \varepsilon(xy)$. The ideal of orphans is given by:

$$O := \ker \theta_R^\vee = \{y \in K_R \mid \forall x \in R, \ \varepsilon_R(xy) = 0\} \subset K_R. \tag{4.15}$$

Let $k \geq 0$ be an integer. Orphans are said to be k-semi-acyclic if $H^i(O) = 0$ for $n/2 + 1 \leq i \leq k$, where $n = \dim M$. Note that this condition is empty if $k = n/2$. Moreover, Poincaré–Lefschetz duality in cohomology implies that if O is $(n + 1)$-semi-acyclic, then it is fully acyclic.

Let us work by induction and assume that O is $(k - 1)$-semi-acyclic for $n/2 \leq k - 1 < n + 1$. Let us show that we can replace (R, R_∂) by a new model whose orphans are k-semi-acyclic. The goal is to build an extension of the exact short sequence:

$$
\begin{array}{ccccccccc}
0 & \longrightarrow & K_R & \lhook\joinrel\longrightarrow & R & \overset{\rho}{\relbar\joinrel\twoheadrightarrow} & R_\partial & \longrightarrow & 0 \\
 & & \downarrow & & \downarrow & & \| & & \\
0 & \longrightarrow & \hat{K}_R & \lhook\joinrel\longrightarrow & \hat{R} & \overset{\hat{\rho}}{\relbar\joinrel\twoheadrightarrow} & R_\partial & \longrightarrow & 0
\end{array}
\tag{4.16}
$$

as well as an extension $\hat{\varepsilon}$ of ε (which still satisfies Stokes' formula) so that the orphans of $(\hat{R}, \hat{R}_\partial)$ are k-semi-acyclic. Going up to $k = n + 1$ will allow us to conclude the proof.

Let $l = \dim(O^k \cap \ker d) - \dim(d(O^{k-1}))$ and choose cycles $\alpha_1, \ldots, \alpha_l \in O^k$ which form a basis of a complementary subspace of $d(O^{k-1})$ in $\ker d$, i.e.:

$$O^k \cap \ker d = d(O^{k-1}) \oplus \langle \alpha_1, \ldots, \alpha_l \rangle. \tag{4.17}$$

These cycles are the obstructions to the k-semi-acyclicity of O. Note that by definition, $\theta_R^\vee(\alpha_i) = 0$. As $\theta_R^\vee$ is a quasi-isomorphism (the pair (R, R_∂) verifies Poincaré–Lefschetz duality in cohomology), we can thus find $\gamma_i' \in K_R^{k-1}$ such that $d\gamma_i' = \alpha_i$.

Let m be the total dimension of $H^*(R) = H^*(M)$, which is (by duality) also the total dimension of $H^*(K_R) = H^*(M, \partial M)$. Choose cycles $h_1, \ldots, h_m \in R$ which form a basis in cohomology. By duality, we can find $h_1', \ldots, h_m' \in K_R$ which form a basis in cohomology and such that $\varepsilon(h_i h_j') = \delta_{ij}$. We then define:

$$\gamma_i := \gamma_i' - \sum_j \varepsilon(\gamma_j' h_j) h_j' \in K^{k-1}. \tag{4.18}$$

A small calculation shows that $d\gamma_i = d\gamma_i' = \alpha_i$ and that for any cycle $y \in R$, $\varepsilon(\gamma_i y) = 0$. We can then extend R by defining:

$$\hat{R} := \big(R \otimes S(\underbrace{c_1, \ldots, c_l}_{\deg = k-1}, \underbrace{w_1, \ldots, w_l}_{\deg = k-2}), dc_i = \alpha_i, dw_i = c_i - \gamma_i \big). \tag{4.19}$$

We can check easily that the inclusion $R \subset \hat{R}$ is a quasi-isomorphism, and that ρ extends into $\hat{\rho} : \hat{R} \to R_\partial$ by $\hat{\rho}(c_i) = \hat{\rho}(w_i) = 0$. Let T be a linear complement of $d(R)$ in R. Then ε extends (where $x \in A$ and $t \in T$) with:

- $\hat{\varepsilon}(x) = \varepsilon(x)$;
- $\hat{\varepsilon}(w_i dx) = (-1)^k \varepsilon(\gamma_i x)$;
- $\hat{\varepsilon}(c_i c_j) = -\varepsilon(\gamma_i \gamma_j)$;
- $\hat{\varepsilon}(w_i) = \hat{\varepsilon}(w_i t) = \hat{\varepsilon}(c_i) = \hat{\varepsilon}(c_i x) = \hat{\varepsilon}(c_i c_j x) = \hat{\varepsilon}(c_i w_j) = \hat{\varepsilon}(c_i w_j x) = \hat{\varepsilon}(w_i w_j) = \hat{\varepsilon}(w_i w_j x) = 0$,
- $\hat{\varepsilon}$ vanishes on elements of degree different from n.

It only remains to show that $(\hat{R}, R_\partial)$ is such that its ideal of orphans is k-semi-acyclic, which is an easy exercise. $\qquad\Box$

Remark 4.13 The proof above follows pretty closely the proof given by Lambrechts and Stanley [LS08b] for the boundary-free case. Recently, a different approach based on Hodge decompositions has been developed by Hájek [Háj20].

In order to define the Lambrechts–Stanley model, we needed diagonal classes. We also have an analog of the diagonal class in PLD pair.

Definition 4.14 Let (B, B_∂, λ) be a PLD pair, $K = \ker \lambda$, I be the orphans, and $P = B/I$. If $\{x_i\}$ is a graded basis of K and $\{x_i^\vee\}$ is the dual basis of P, then we define a cocycle of degree n by:

$$\Delta_{KP} := \sum_i (-1)^{\deg x_i} x_i \otimes x_i^\vee \in K \otimes P. \tag{4.20}$$

For simplicity we also define Δ_P as the image of Δ_{KP} by the map $K \otimes P \hookrightarrow B \otimes P \twoheadrightarrow P \otimes P$.

One can also interpret Δ_P by dualizing the multiplication $K \otimes K \to K$ into a coproduct $P[-n] \to P[-n] \otimes P[-n]$. Then Δ_P is the image of $1 \in P$ under this coproduct. In particular, we have the following property, which is easily checked (cf. Lemma 3.12):

Proposition 4.15 *Let (B, B_∂, λ) be a PLD pair, I be the orphans, and $P = B/I$. Then, for any $x \in P$:*

$$(x \otimes 1)\Delta_P = (1 \otimes x)\Delta_P. \tag{4.21}$$

We will also need a particular element, the "section". Let $s : B_\partial \to B$ be a (linear) section of λ, which is of course generally not an algebra morphism nor a chain map. However, we have $s(dx) - ds(x) \in K$ for all $x \in B_\partial$ (because $\lambda(s(dx)) = dx = d\lambda(s(x)) = \lambda(ds(x))$).

Definition 4.16 Let (B, B_∂, λ) be a PLD pair and let $s : B_\partial \to B$ be a linear section of λ. The element $s \in B \otimes B_\partial^\vee$ corresponds to an element $\sigma_B \in B \otimes B_\partial$ (of degree $n - 1$) by the Poincaré duality of B_∂ and is called the *section* of the PLD pair.

The following proposition is also a short exercise:

Proposition 4.17 *Let (B, B_∂, λ) be a PLD pair, $K = \ker \lambda$, I be the orphans, and $P = B/I$. Then $d\sigma_B$ belongs to $K \otimes B_\partial$, and we have $(\mathrm{id} \otimes \pi)(d\sigma_B) = \Delta_{KP}$.*

Example 4.18 Let us consider again Example 4.10. In this example, the diagonal class is $\Delta_{KP} = w \otimes 1$ and the only possible choice of σ_B is $1 \otimes v + v \otimes 1$. The differential of σ_B is $1 \otimes w + w \otimes 1$ which projects to Δ_{KP} via $\mathrm{id} \otimes \pi : K \otimes B \to K \otimes P$.

4.3   Graphical Models

Let M be a compact manifold with boundary $\partial M = N$. Our objective, in the section, is to define a model for the monoid $\mathrm{Conf}_{N \times \mathbb{R}}$ and its module Conf_M in terms of graph complexes (as in Sect. 3.4). For reasons of contravariance, the model of $\mathrm{Conf}_{N \times \mathbb{R}}$ will be a comonoid in the category of symmetric collections of CDGA (see Chap. 5 for a precise definition), and the model of Conf_M will be a comodule over this comonoid. We will start by describing compactifications of $\mathsf{Conf}_{N \times \mathbb{R}}$ and Conf_M inspired by the Fulton–MacPherson compactifications. We will then define propagators on these compactifications, and then the aforementioned graph complexes together with their algebraic structures. We will then explain how to simplify these graph complexes up to homotopy and finally prove that they are real models of the configuration spaces that we consider.

4.3.1 Compactifications

Let us first deal with compactifications of configuration spaces of collars around boundaries. Let N be a compact manifold without boundary, of dimension $n - 1$. Informally, we will think of N as the boundary of an n-dimensional manifold.

There is a natural action of $\mathbb{R}_{>0}$ on $\mathsf{Conf}_{N \times \mathbb{R}_{>0}}$ by multiplication on the second factor. We will define a compactification $\mathsf{aFM}_N(U)$ (the symbol "a" is for "algebra") of the space $\mathsf{Conf}_{N \times \mathbb{R}_{>0}}(U)/\mathbb{R}_{>0}$ such that the aFM_N collection forms a strictly associative and unitary monoid in the category of symmetric sequences of topological spaces. Assume that N is embedded as an SA submanifold of $\mathbb{R}^D$, for D large enough. Then we define several maps:

- For $i \in U$,

$$p_i : \mathsf{Conf}_{N \times \mathbb{R}_{>0}}(U) \to N \tag{4.22}$$

  is the projection of a configuration in $N \times \mathbb{R}_{>0}$ to its ith coordinate, followed by the projection onto N.

- Let us view the sphere $\mathbb{S}^D$ as the quotient space $((\mathbb{R}^D \times \mathbb{R}) \setminus \{0\})/\mathbb{R}_{>0}$. In the next formula, we denote the elements of that space as equivalence classes of the form $[u : t]$ for $u \in \mathbb{R}^D$ and $t \in \mathbb{R}$. Then we define, for $i \neq j$, a map $\theta_{ij} : \mathsf{Conf}_{N \times \mathbb{R}_{>0}}(U) \to \mathbb{S}^D$ on an element $(\underline{x}, \underline{h}) = (x_i, h_i)_{i \in U} \in \mathsf{Conf}_{N \times \mathbb{R}_{>0}}(U)$ by:

$$\theta_{ij}(\underline{x}, \underline{h}) := \left[x_j - x_i : \frac{h_j}{h_i} - 1 \right]. \tag{4.23}$$

- Finally, for $i \neq j \neq k \neq i$, we define $\delta_{ijk} : \mathsf{Conf}_{N \times \mathbb{R}_{>0}}(U) \to S^{3D+2}$ by:

$$\delta_{ijk}(\underline{x}, \underline{h}) := [\alpha_{ij} : \alpha_{jk} : \alpha_{ik}], \tag{4.24}$$

where we define the vectors $\alpha_{ij} = (x_j - x_i) \oplus (h_j/h_i - 1)$ as (nonzero) elements of $\mathbb{R}^D \oplus \mathbb{R} \cong \mathbb{R}^{D+1}$, which we then project to the unit sphere.

These maps are clearly invariant under the action of $\mathbb{R}_{>0}$. Together, they define an embedding of the space $\mathsf{Conf}_{N \times \mathbb{R}_{>0}}(U)/\mathbb{R}_{>0}$ into the product:

$$N^U \times \left(\mathbb{S}^D\right)^{\mathsf{Conf}_U(2)} \times ([0, +\infty] \times [0, +\infty])^{\mathsf{Conf}_U(3)}. \tag{4.25}$$

Definition 4.19 The space $\mathsf{aFM}_N(U)$ is the closure of the embedding defined by the maps (4.22), (4.23) and (4.24).

The following proposition can be checked in the same way as it can be checked for the Fulton–MacPherson compactification FM_M in Chap. 3.

Proposition 4.20 *The space* $\mathsf{aFM}_N(U)$ *is a compact SA manifold with boundary of dimension* $n\#U - 1$, *with interior* $\mathsf{Conf}_{N \times \mathbb{R}_{>0}}(U)/\mathbb{R}_{>0}$.

Example 4.21 It is instructive to study the case where $U = \{1, 2\}$ has two elements. We will moreover need an explicit description of the compactification in that case, in order to define the propagator in Sect. 4.3.2. The space $\mathsf{aFM}_N(U)$ then has four "strata", which we will number from I to IV (see Fig. 4.4).

  I. (codimension 0) The interior of the manifold, which is homeomorphic to $\mathsf{Conf}_{N \times \mathbb{R}_{>0}}(U)/\mathbb{R}_{>0}$. It contains the "classical" configurations, i.e., the one where points have not collided.
 II. (codimension 1) The set of configurations $((x_1, h_1), (x_2, h_2))$ such that $h_2/h_1 = \infty$, that is, the second point is infinitesimally close to $N \times \{\infty\}$ (and thus the first is infinitesimally close to $N \times \{0\}$). This stratum is homeomorphic to $N \times N$.

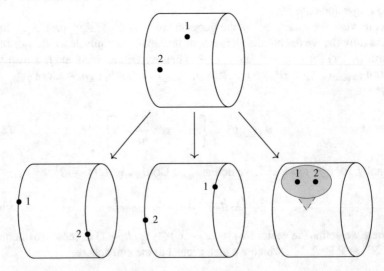

Fig. 4.4 Strata of $\mathsf{aFM}_N(2)$, where $N = \mathbb{S}^1$. From top to bottom and left to right: I, II, III, IV. The stratum at the target of an arrow is included in the boundary of the stratum at the source of an arrow (e.g., stratum II is included in the boundary of stratum I)

III. (codimension 1) The set of configurations $((x_1, h_1), (x_2, h_2))$ such that $h_1/h_2 = +\infty$, that is, the first point is infinitesimally close to $N \times \{\infty\}$ (and thus the second is infinitesimally close to $N \times \{0\}$). This stratum is also homeomorphic to $N \times N$.

IV. (codimension 1) The set of configurations $((x_1, h_1), (x_2, h_2))$ such that $(x_1, h_1) \approx (x_2, h_2)$, i.e. the two points are infinitesimally close to each other. The two points are somewhere in $N \times \mathbb{R}_{>0}$, but because of the quotient by the action of $\mathbb{R}_{>0}$, the only remaining information is their position in N. This stratum is then homeomorphic to the normal fiber bundle of $\Delta_N \times \{1\} \subset N^2 \times \mathbb{R}_{>0}$, where $\Delta_N \subset N^2$ is the diagonal.

Proposition 4.22 *The collection of all the spaces* aFM_N *forms a monoid, i.e., given two finite sets U and V, we have a map:*

$$\mathsf{aFM}_N(U) \times \mathsf{aFM}_N(V) \to \mathsf{aFM}_N(U \sqcup V) \tag{4.26}$$

and the collection of all such maps satisfies associativity and unitality conditions.

Let us now explain the monoid structure. The map of Eq. (4.26) is obtained by gluing the component $N \times \{\infty\}$ of the first configuration with the $N \times \{0\}$ component of the second. This can be defined explicitly in terms of the coordinates $(p_i, \theta_{ij}, \delta_{ijk})$ by considering that the points of the second configuration are infinitely far away from the points of the first one (with $h_i/h_j = \infty$ if $i \in U$ and $j \in V$). This is easily seen to be a strictly associative product, with a unit given by the empty configuration $\varnothing \in \mathsf{aFM}_N(\varnothing)$.

Let us now turn to compactification of configuration spaces of manifolds with boundary. Suppose that $\partial M = N$. We can define a compactification $\mathsf{mFM}_M(U)$ of $\mathsf{Conf}_M(U)$ which is a right module over the monoid aFM_N (the symbol "m" is for "module"). In the interior of M, the compactification is constructed as the compactification FM_M of Sect. 3.2. On the boundary, it is constructed by gluing aFM_N along a collar $N \times \mathbb{R}_{\geq 0} \hookrightarrow M$. We then get the following results:

Proposition 4.23 *The spaces* $\mathsf{mFM}_M(U)$ *are compact SA manifolds with boundary of dimension $n\#U$, with interior* $\mathsf{Conf}_{\mathring{M}}(U)$.

Proposition 4.24 *The collection of all the spaces* $\mathsf{mFM}_M(U)$ *forms a right module over the monoid aFM_N, i.e., given two finite sets U and V, we have maps*

$$\mathsf{mFM}_M(U) \times \mathsf{aFM}_N(V) \to \mathsf{mFM}_M(U \sqcup V) \tag{4.27}$$

and the collection of all such maps satisfies associativity and unitality conditions.

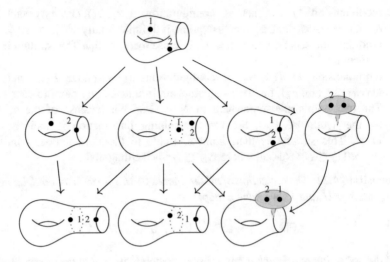

Fig. 4.5 Strata of $\mathsf{mFM}_M(2)$. From top to bottom and left to right: I, II, V, III, IV, VI, VII, VIII

Example 4.25 Just as in Example 4.21, it is interesting to describe the decomposition in strata of the manifold $\mathsf{mFM}_M(\{1, 2\})$. The space $\mathsf{mFM}_M(\{1, 2\})$ has eight strata, numbered from I to VIII (see Fig. 4.5):

  I. (codimension 0) The interior of the manifold, which is homeomorphic to $\mathsf{Conf}_{\mathring{M}}(2)$. The two points remain within M and stay distant from each other.
 II. (codimension 1) The configurations where the second point is infinitesimally close to the boundary. This stratum is homeomorphic to $\mathring{M} \times \partial M$.
III. (codimension 1) The configurations where the first point is infinitesimally close to the boundary. This stratum is homeomorphic to $\partial M \times \mathring{M}$;
 IV. (codimension 1) The configurations where the two points are infinitesimally close to each other but far from the boundary. This stratum is homeomorphic to $\partial \mathsf{FM}_{\mathring{M}}(2)$;
  V. (codimension 1) The configurations where the two points are infinitesimally close to the boundary, but far from each other. This stratum is homeomorphic to $\mathsf{Conf}_{N \times \mathbb{R}}(2)/\mathbb{R}_{>0}$, i.e., stratum I of $\mathsf{aFM}_M(2)$.
 VI. (codimension 2) The configurations where the two points are infinitesimally close to the boundary and the second is infinitely closer to $N \times \{\infty\}$ than the first. This stratum is homeomorphic to $N \times N$, i.e., stratum II of $\mathsf{aFM}_N(2)$.
VII. (codimension 2) The inverse situation of the previous one. This stratum is homeomorphic to $N \times N$, i.e., stratum III of $\mathsf{aFM}_N(2)$.
VIII. (codimension 2) The configuration where the two points are infinitesimally close to the boundary and moreover infinitesimally close to each other. This stratum is homeomorphic to stratum IV of $\mathsf{aFM}_N(2)$.

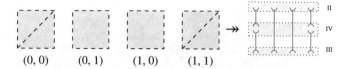

Fig. 4.6 The space $\mathsf{Conf}_{N \times \mathbb{R}_{>0}}(2)$ (in black), its quotient by $\mathbb{R}_{>0}$ (in red) and the strata of the boundary of $\mathsf{aFM}_N(2)$, for $N = \{0, 1\}$

Example 4.26 Generally speaking, it is not easy to represent the spaces $\mathsf{aFM}_N(U)$ and $\mathsf{mFM}_M(U)$. Indeed, their dimension increases rapidly: $\dim \mathsf{aFM}_N(U) = n\#U - 1$ and $\dim \mathsf{mFM}_M(U) = n\#U$. However, let us look at the case where $M = [0, 1]$ and $N = \partial M = \{0, 1\}$.

We will draw spaces homeomorphic to $\mathsf{aFM}_N(2)$ and $\mathsf{mFM}_M(2)$ to illustrate how the previously defined strata intersect. The drawings will of course not be completely faithful to reality: we project subspaces of $\mathbb{R}^4$ on a two-dimensional plane.

Let us first describe $\mathsf{aFM}_N(2)$ for $N = \{0, 1\}$. The space $N \times \mathbb{R}_{>0}$ is a disjoint union of two open half-lines. Its square, $(N \times \mathbb{R}_{>0})^2$, is a disjoint union of four open quarter planes, indexed by $N^2 = \{(0, 0), (0, 1), (1, 0), (1, 1)\}$. To obtain the space $\mathsf{Conf}_{N \times \mathbb{R}_{>0}}(2) \subset (N \times \mathbb{R}_{>0})^2$, one must remove the diagonals of the quarter planes indexed by $(0, 0)$ and $(1, 1)$. Graphically, this give Fig. 4.6, where the dotted lines represent missing parts of the spaces (which are supposed to extend to infinity in two directions).

The interior of $\mathsf{aFM}_N(2)$ is the quotient of $\mathsf{Conf}_{N \times \mathbb{R}_{>0}}(2)$ by the action of $\mathbb{R}_{>0}$ on the second factors, i.e. $\lambda \cdot ((x, t), (x', t')) = ((x, \lambda t), (x', \lambda t'))$. In our case, this space is the disjoint union of six open intervals, represented for example by configurations of the type $((x, 1), (x', t'))$. This is represented by the red part of Fig. 4.6.

The compactification $\mathsf{aFM}_N(2)$ is obtained by adding boundary components to $\mathsf{Conf}_{N \times \mathbb{R}_{>0}}(2)/\mathbb{R}_{>0}$. Here, it is necessary to add the two extremities of each of the six intervals which compose $\mathsf{Conf}_{N \times \mathbb{R}_{>0}}(2)$, for a total of twelve points. Stratum II corresponds to the case where the second point goes to infinity ($t'/t \to +\infty$), stratum III to the case where the first point goes to infinity ($t'/t \to 0$), and stratum IV to the case where the two points converge towards the same position ($x = x'$ and $t/t' \to 1$).

Let us now describe $\mathsf{mFM}_M(2)$ for $M = [0, 1]$. Its interior, $\mathsf{Conf}_{\mathring{M}}(2)$, is an open square with the diagonal removed. Its boundary $\partial \mathsf{mFM}_M(2)$ is itself a manifold with boundary. The inside of the boundary consists of nine intervals:

- the four sides of the square: two for stratum II (the first point converges towards the boundary) and two for stratum III (the second point converges towards the boundary);
- two on each "side" of the diagonal, forming stratum IV;

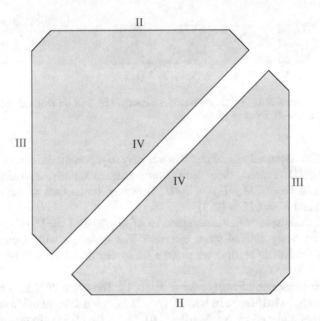

Fig. 4.7 The space $mFM_{[0,1]}(2)$ and its one-codimensional strata. The small segments (in black) are all part of the stratum V. The strata VI, VII, and VIII are the corners

- four intervals which form stratum V (the two points tend towards the boundary at different speeds so as not to collide) and which connect the preceding intervals according to combinatorics encoded in Fig. 4.7.

We have represented a space homeomorphic to $mFM_M(2)$, together with its decomposition in strata, in Fig. 4.7. For the purpose of visualization, we have represented the two ends of the segments that make up the stratum V as being different point in the two-dimensional place. They are actually both projected onto the corners of the square $M^2 = [0,1]^2$, and the segment that joins them is orthogonal M^2. In the same way, the two segments of the diagonal (stratum IV) are projected onto the diagonal $\Delta_M = \{(x,x) \mid x \in M\} \subset M^2$.

4.3.2 Propagators

Let $(M, \partial M = N)$ be a manifold with boundary and let $n = \dim M$. In this section, we will define the propagators, i.e., the forms on $aFM_N(2)$ and $mFM_M(2)$, that will allow us to define maps from graph complexes to the CDGAs of forms on the compactifications.

Suppose that the pair (M, N) has a Poincaré–Lefschetz duality model as in Sect. 4.2:

$$
\begin{array}{ccccccc}
& & K := \ker \lambda & & & & \\
& \nearrow^{\text{accoupl.}}_{\text{non dégén.}} & \downarrow & & & & \\
P := B/\ker\theta_B & \xleftarrow{\;\;\sim\;\;}_{\pi} & B & \xleftarrow{\;\;\sim\;\;}_{f} & R & \xrightarrow{\;\;\sim\;\;}_{g} & \Omega^*_{\mathrm{PA}}(M) \\
& & \downarrow{\lambda} & & \downarrow{\rho} & & \downarrow{\text{res}} \\
& & B_\partial & \xleftarrow{\;\;\sim\;\;}_{f_\partial} & R_\partial & \xrightarrow{\;\;\sim\;\;}_{g_\partial} & \Omega^*_{\mathrm{PA}}(\partial M)
\end{array}
\qquad (4.28)
$$

Remark 4.27 If $(M, \partial M)$ does not admit a Poincaré–Lefschetz duality model, it will still be possible to define graph complexes, by decorating the vertices by elements of $S(\tilde{H}^*(M) \oplus H^*(M, \partial M))$ and $S(\tilde{H}^*(\partial M))$ instead of B and B_∂, see [CILW18] for details.

Let us denote the diagonal class by $\Delta_{KP} \in K \otimes P$, and let $\Delta_P \in P \otimes P$ be its image in $P \otimes P$ and $\sigma_B \in B \otimes B_\partial$ the "section" (which satisfies $(\pi \otimes \mathrm{id})(d\sigma_B) = (1 \otimes \lambda)(\Delta_{KP})$), see Definitions 4.14 and 4.16. Thanks to rather simple arguments, we can lift Δ_P and σ_B into elements $\Delta_R \in R \otimes R$ and $\sigma_R \in R_\partial \otimes R$ satisfying $d\sigma_R \in R_\partial \otimes \ker \rho$. From now on, we fix these lifts.

Let us first define the propagator on $\mathsf{aFM}_N(2)$.

Proposition 4.28 *There is a form* $\varphi_\partial \in \Omega^{n-1}_{\mathrm{PA}}(\mathsf{aFM}_N(2))$, *called the* propagator, *such that:*

- *the form is closed, i.e., $d\varphi_\partial = 0$;*
- *the restriction of φ_∂ to stratum IV of $\mathsf{aFM}_N(2)$ (i.e., the two points are infinitesimally close, $x_1 \approx x_2$) is a global angular form;*
- *the restriction of φ_∂ to stratum II of $\mathsf{aFM}_N(2)$ (i.e., the second point goes to infinity, $x_2 \to \infty$) is equal to the image of $\sigma_\partial := (\mathrm{id} \otimes \rho)(\sigma_R)$;*
- *the restriction of φ_∂ to stratum III of $\mathsf{aFM}_N(2)$ (i.e., the first point goes to infinity, $x_1 \to \infty$) vanishes;*
- *for any element $\alpha \in R_\partial$, we have $\int_y \varphi_\partial(x, y)\alpha(y) = 0$.*

Proof The proof is similar to the one of Proposition 3.88. For the sake of this argument, we will define the propagator on $\mathsf{Conf}_{N \times [0,1]}$, and then apply a rescaling of the form $t \mapsto t/(t - 1)$ on the second coordinate. Let ψ be a global angular form on the fiber bundle over N defined by stratum IV. It extends into a form whose support is contained in a neighborhood of this stratum. Its differential $d\psi$ is closed (since $d^2 = 0$) and its support is disjoint from stratum IV and the diagonal. It thus represents a class in:

$$
H^*(N \times N \times [0, 1], N \times N \times \{0, 1\}) \cong (H^*(M) \otimes H^*(M))[-1]. \qquad (4.29)
$$

We check with Stokes' formula that $d\psi \pm \sigma_\partial \wedge d(1-t)$ is the boundary of a form γ on $M \times M \times [0,1]$. We thus set $\psi' := \psi - \gamma \pm \sigma_\partial \wedge (1-t)$. The form γ vanishes on the boundary and thus does not change the behavior on the three strata. Moreover, the other two forms give exactly what we wanted on strata II and III, where ψ vanishes. Finally we replace ψ' by:

$$\varphi_\partial := \psi \pm \int_3 \psi'_{13}(\sigma_\partial)_{32}dt \pm \int_3 \psi'_{23}(\sigma_\partial)_{31}dt \mp \int_{3,4} \psi'_{34}(\sigma_\partial)_{23}(\sigma_\partial)_{14}, \qquad (4.30)$$

where $\int_i$ (resp. $\int_{i,j}$) is the integral along the fibers of the projection that forgets the ith coordinate (resp. the ith and the jth coordinates), and ω_{ij} (for some form ω) is the pullback of ω along the projection that keeps only the ith and jth coordinates. The last property then follows from an explicit computation.   □

We can moreover define a propagator on $\mathsf{mFM}_M(2)$ in a similar manner:

Proposition 4.29 *There is a form $\varphi \in \Omega_{\mathrm{PA}}^{n-1}(\mathsf{mFM}_M(2))$, also called the* propaga-*tor, such that:*

- *Its differential is such that $d\varphi = \Delta_R$;*
- *On strata II ($x_2 \to \partial M$), III ($x_1 \to \partial M$) and V ($x_1, x_2 \to \partial M$), we have respectively:*

$$\varphi|_{II} = \sigma_R, \qquad \varphi|_{III} = 0, \qquad \varphi|_V = \varphi_\partial. \qquad (4.31)$$

- *on stratum IV ($x_1 \approx x_2$), φ is a global angular form;*
- *for all $\alpha \in R$, we have $\int_y \varphi(x,y)\alpha(y) = 0$.*

Proof The proof is similar to the one of Proposition 4.28. We just take into account more strata. Note that on the two-codimensional strata VI, VII, and VIII, the behavior is that of $\phi|_V = \varphi_\partial$ and it is consistent with Proposition 4.28.   □

4.3.3 Graph Complexes

We will now define real models of aFM_N and mFM_M. These models will be given by graph complexes, in the line of Sect. 3.4.

Definition 4.30 Let U be a finite set. The graph complex $\mathsf{aGraphs}_{R_\partial}(U)$ is spanned by graphs of the following type:

- the graph has a set of vertices called "external", in bijection with U, and an arbitrary finite set I of vertices called "internal";
- internal vertices are made undistinguishable using the action of the symmetric group;
- each vertex is decorated by an element of R_∂;
- the edges are directed;

- the edges are of degree $n - 1$, the internal vertices of degree $-n$, and the decorations have the same degree as in R_∂;
- if n is even, then the set of all edges is equipped with an order, but if two graphs differ only in the ordering of their edges, then they are identified up to a sign (determined by the Koszul rule of signs);
- the integration procedure described in Sect. 3.4.4 allows us to define a number $w(\gamma)$—the *partition function* of $N \times \mathbb{R}_{>0}$ on γ—associated to a connected graph γ containing only internal vertices, and we set the relation $\Gamma \sqcup \gamma \equiv w(\gamma) \cdot \Gamma$ for an arbitrary graph $\Gamma \in \mathsf{aGraphs}_{R_\partial}(U)$.

The product on the CDGA $\mathsf{aGraphs}_{R_\partial}$ consists in gluing two graphs together along their external vertices. Its differential is the sum of two terms:

- the internal differential d_{R_∂} which acts on the decorations as a derivation;
- the contracting differential d_{contr}, which is the sum over all the ways to contract an edge of the graph and multiplying the decorations of its extremities.

Finally, the coproduct $\Delta : \mathsf{aGraphs}_{R_\partial}(U \sqcup V) \to \mathsf{aGraphs}_{R_\partial}(U) \otimes \mathsf{aGraphs}_{R_\partial}(V)$, which reflects the monoid structure of the collection aFM_N, is given as follows. It consists in cutting the graph in two components, with U on the left, V on the right, and summing on all the possible ways to distribute the internal vertices on the left and on the right. We then replace the edges that connect vertices of two different components by σ_∂ if they go from left to right, and by 0 otherwise. See Fig. 4.8 for an example.

Proposition 4.31 *There are morphisms* $\mathsf{aGraphs}_{R_\partial}(U) \to \Omega^*_{\text{PA}}(\mathsf{aFM}_N(U))$ *for all finite sets U which are compatible with the monoid structure of* aFM_N.

Proof Very similar to that of Sect. 3.4. $\square$

Definition 4.32 The graph complex $\mathsf{mGraphs}_R(U)$ is spanned by graphs of the following type:

- the graph has a set of external vertices, in bijection with U, and an arbitrary finite set I of internal vertices;
- using the action of the symmetric group, internal vertices are made undistinguishable;
- each vertex is decorated by an element of R;
- each edge is directed;

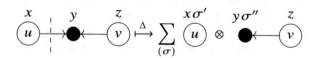

Fig. 4.8 The coproduct $\mathsf{aGraphs}_{R_\partial}(\{u, v\}) \to \mathsf{aGraphs}_{R_\partial}(\{u\}) \otimes \mathsf{aGraphs}_{R_\partial}(\{v\})$, where $x, y, z \in R_\partial$ and $\sigma = \sum_{(\sigma)} \sigma' \otimes \sigma''$

- the edges are of degree $n - 1$, the internal vertices of degree $-n$, and the decorations have the same degree as in R;
- if n is even, then the set of all edges is equipped with an order, but if two graphs differ only in the ordering of their edges, then they are identified up to a sign (determined by the Koszul rule of signs);
- the integration procedure described in Sect. 3.4.4 allows to define a number $W(\gamma)$—the *partition function* of M on γ—associated with a connected graph γ containing only internal vertices, and we set the relation $\Gamma \sqcup \gamma \equiv W(\gamma)\Gamma$ for an arbitrary $\Gamma \in \mathsf{mGraphs}_R(U)$.

The product on $\mathsf{mGraphs}_R$ consists of pasting two graphs together along their external vertices. The coaction $\Delta : \mathsf{mGraphs}_R(U \sqcup V) \to \mathsf{mGraphs}_R(U) \otimes \mathsf{aGraphs}_{R_\partial}(V)$ that reflects the module structure of mFM_M over aFM_N is defined as follows. It consists in cutting the graph in two parts, with U on the left, V on the right, and summing over all the possible ways to distribute the internal vertices on the left and on the right. We then replace the edges that connect vertices of two different parts by σ if they go from left to right, and by 0 if not. Finally, we apply $\rho : R \to R_\partial$ on all the decorations of the graph on the right. See Fig. 4.8, where we would apply ρ to the decorations of the vertices of the second tensor factor.

Finally, the differential is the sum of four terms:

- the internal differential d_R, which acts on the decorations as a derivation;
- the cutting differential d_{split}, which is the sum of all the ways to cut an edge and multiply the decoration of its extremities by Δ_R;
- the contracting differential d_{contr}, which is the sum over all the ways of contracting an edge in the graph and multiplying the decoration of its extremities;
- the twist by w, defined similarly to the coaction Δ for $V = \varnothing$, i.e., we cut off graphs consisting only of internal vertices, and we apply the partition function w to the part that has been cut off.

Proposition 4.33 *There are morphisms* $\mathsf{mGraphs}_R(U) \to \Omega^*_{\mathrm{PA}}(\mathsf{mFM}_M(U))$ *for all finite sets U which are compatible with the coaction of* aFM_N.

4.3.4 Simplification of the Partition Functions

The definitions of the above graph complexes depend on integrals, the partition functions w and W, that are difficult to compute. As in Sect. 3.4.6, we would like to simplify the expressions of w and W. Unfortunately, there is no argument as simple as in the case without a boundary to show that these functions are trivial up to homotopy. More sophisticated techniques, that we are now going to explain, are needed.

4.3.4.1   Partition Function of the Cylinder on the Boundary

As in Sect. 3.4.5, we can define a differential graded algebras in which our partition functions live.

Definition 4.34 Let aGC_{R_∂} be the dg-module spanned by connected graphs of type $aGraphs_{R_\partial}$ that only contain internal vertices, and without modding out by the relations of the last item of Definition 4.30 (i.e., $\gamma \equiv w(\gamma)$).

The free graded algebra on aGC_{R_∂} is equipped with a differential that has a description similar to the one of $aGraphs_{R_\partial}$. The element w mentioned in Definition 4.30 defines a morphism of CDGAs from this quasi-free algebra to $\mathbb{R}$. We have the following point of view, just like in Chap. 3:

Proposition 4.35 *The dg-module* $aGC_{R_\partial}^\vee[-1]$ *is equipped with a dg-Lie algebra structure such that morphisms of CDGAs* $(S(aGC_{R_\partial}), d) \to \mathbb{R}$ *are in bijection with Maurer–Cartan elements of* $aGC_{R_\partial}^\vee[-1]$.

Proof This follows from the general arguments of Sect. 3.4.5. As in Sect. 3.4.5, the differential of $aGraphs_{R_\partial}$ is quadratic-linear in terms of the elements of aGC_R. It thus induces a differential and a Lie bracket on the dual space $aGC_{R_\partial}^\vee[-1]$. The differential of $aGC_{R_\partial}^\vee[-1]$ is the dual of the linear part, while the Lie bracket is the dual of the quadratic part. Concretely, the differential is a sum of two terms: the internal differential, which acts on decorations, and the connecting differential, which connects two vertices in all possible ways (and precomposes decorations by multiplication with Δ_{R_∂}). The Lie bracket is the sum of all possible ways of connecting two graphs by an edge (de-multiplying decorations too).   □

Applied to the present case, this correspondence tells us that the morphism w corresponds to a Maurer–Cartan element $w \in aGC_{R_\partial}^\vee[-1]$. In the dual basis of the graphs basis, one can write $w = \sum_\Gamma w(\Gamma)\Gamma$. This point of view allows us to apply classical theorems about Maurer–Cartan elements. Let us now simplify w up to homotopy.

Definition 4.36 Denote by $w_0 \in aGC_{R_\partial}^\vee[-1]$ the *partition function*, i.e., the restriction of w to singleton graphs (cf. Definition 3.101).

Proposition 4.37 *The Maurer–Cartan elements w and w_0 are homotopic.*

Proof This is done in several steps. Note that in the following discussion, we take the convention that the valence of a vertex is increased by 1 if the decoration of this vertex is of positive degree.

We start by showing that $aGC_{R_\partial}^\vee$ is quasi-isomorphic to its Lie sub-dg-algebra $aGC_{R_\partial}^{\geq 2,\vee}$ generated by graphs whose vertices are at least bivalent. This results from a classical argument in the theory of graph complexes (cf. [Wil16], compare with the discussion after Definition 3.109). The element $w - w_0$ is therefore homotopic to an element of $aGC_{R_\partial}^{\geq 2,\vee}$.

We then show that the cohomology of $\mathrm{aGC}_{R_\partial}^{\geq 2,\vee}$ differs from the cohomology of $\mathrm{aGC}_{R_\partial}^{\geq 3,\vee}$ only for those graphs with one loop (i.e. the loop-shaped graphs). We moreover show that this one-loop part is generated by the circular graphs as defined in Sect. 3.4.6. This results from another classical argument, counting the bivalent vertices. All circular graphs are of negative degree, so $w - w_0$ vanishes on them. It follows that $w - w_0$ is homotopic to an element of $\mathrm{aGC}_{R_\partial}^{\geq 3,\vee}$.

Finally, a simple degree counting argument shows that the cohomology of $\mathrm{aGC}_{R_\partial}^{\geq 3,\vee}$ vanishes in degree 1 if $\dim M = n$ is at least 3, or equivalently, $\dim N \geq 2$. Indeed, let $\Gamma \in \mathrm{aGC}_{R_\partial}^{\geq 3,\vee}$ be a graph with l loops (i.e. $\dim H^1(\Gamma) = l$ if we see it as a CW-complex of dimension 1). Let us show that $\deg \Gamma \leq -(l-1)(n-3)$. We are going to iteratively find what conditions Γ must satisfy to be of the highest degree possible.

First, if Γ has a (≥ 4)-valent vertex, then we can replace this vertex by an edge and distribute the incident edges (and the possible decoration) arbitrarily between the two new vertices to obtain a graph of higher degree. Pictorially, this corresponds to the following operation:

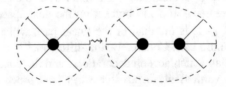

In the worst case, all the vertices of Γ are thus at most trivalent. Since Γ belongs to $\mathrm{aGC}_{R_\partial}^{\geq 3,\vee}$, it follows that all its vertices are exactly trivalent (still with our convention that a decoration of positive degree contributes one to the valence). If Γ has a vertex with two incident edges and a decoration of positive degree, then removing this vertex and replacing it by an edge can only increase the degree. We can therefore assume that all vertices have exactly three incident edges and that all the decorations are of degree 0.

The degree of Γ is therefore $n(2l-2) - (n-1)(3l-3) = -(l-1)(n-3)$, which is negative if $l \geq 2$ and $n \geq 3$. However, the degree of a graph must be equal to 1 for the corresponding integral to be non-zero (because of the quotient by $\mathbb{R}_{>0}$ in the definition of aFM_N). The cases $l = 0$ and $l = 1$ correspond respectively to w_0 and to the loops, which we have already treated.

We conclude from the above that $w - w_0$ is homotopic to zero for $n \geq 3$, so w is homotopic to w_0. For $n = 2$, the boundary N is a union of circles and the result was previously obtained by Willwacher [Wil16]. The result is even clearer if $n = 1$, in which case N is discrete. $\square$

We thus get the following result:

Definition 4.38 We define a new graph complex $\mathsf{aGraphs}^0_{R_\partial}$, exactly as $\mathsf{aGraphs}_{R_\partial}$ except that we use w_0 instead of w to identify an internal component with a number.

Corollary 4.39 *The two comonoids in symmetric collections of CDGAs* $\mathsf{aGraphs}_{R_\partial}$ *and* $\mathsf{aGraphs}^0_{R_\partial}$ *are quasi-isomorphic.*

The model $\mathsf{aGraphs}^0_{R_\partial}$ depends only on the CDGA R_∂, i.e. on the actual homotopy type of N. Once we prove that $\mathsf{aGraphs}_{R_\partial}$ is a model of aFM_N, we can deduce the real homotopic invariance of $\mathsf{Conf}_{N \times \mathbb{R}}$, without any condition on N. This echoes a result of Raptis and Salvatore [RS18], who showed that the homotopy type of $\mathsf{Conf}_{N \times X}(2)$, where X is a contractible space different from the singleton, depends only on the homotopy type of N.

4.3.4.2 Partition Function of the Whole Manifold

We now move on to the simplification of the partition function W. We obtain exactly the same kind of result, but with conditions on M.

Definition 4.40 Let mGC_R be the dg-module spanned by connected graphs of type $\mathsf{mGraphs}_R$ that only contain internal vertices, and without modding out by the relations of the last item of Definition 4.32 (i.e., $\gamma \equiv W(\gamma)$). Its differential is given by the first three summands of the description of Definition 4.32.

The dual dg-module, $\mathsf{mGC}^\vee_R$, is a dg-Lie algebra. The Lie algebra $\mathsf{aGC}^\vee_{R_\partial}[-1]$ acts on $\mathsf{mGC}^\vee_R[-1]$ by derivations, in a way which is dual to the coaction of $\mathsf{aGraphs}$ on $\mathsf{mGraphs}$. Since the element $w \in \mathsf{aGC}^\vee_R[-1]$ is a Maurer–Cartan element, the twisted differential,

$$d_w := d + [w, -] : \mathsf{mGC}^\vee_R[-1] \to \mathsf{mGC}^\vee_R[-1], \tag{4.32}$$

squares to zero, as one can easily check from the Maurer–Cartan equation. One can then check the following, using the Stokes formula:

Proposition 4.41 *The partition function W defines a Maurer–Cartan element in* $(\mathsf{mGC}^\vee_R, d + [w, -])$.

We can verify that, in fact, if we restrict w to the trivial partition function, we still have a Maurer–Cartan element:

Proposition 4.42 *The partition function W is a Maurer–Cartan element in*

$$\mathsf{mGC}^{\vee, w_0}_R := \left(\mathsf{mGC}^\vee_R, d + [w_0, -] \right). \tag{4.33}$$

Definition 4.43 Denote by $W_0 \in \text{mGC}_{R_\partial}^\vee$ the *trivial partition function* (i.e., the restriction of W to singleton graphs), which is a Maurer–Cartan element in $\text{mGC}_R^{\vee, w_0}$.

Proposition 4.44 *If* $\dim M \geq 4$ *and* M *is simply connected, then the Maurer–Cartan elements* W *and* W_0 *are homotopic.*

Proof Our goal is to show that $W - W_0$ is homotopic to zero in the twisted Lie algebra $\text{mGC}_R^{\vee, w_0 + W_0} := (\text{mGC}_R^{\vee, w_0}, [W_0, -])$. The steps are similar to the proof of $w \simeq w_0$ (see Proposition 4.37). There, we show that we can restrict ourselves to (≥ 3)-valent graphs, and mod out the circular graphs. A degree-counting argument similar to the one for $\text{aGC}_{R_\partial}^\vee$ is used to show that the degree of a graph with l loops is at most $-(l-1)(n-3)$. Hence, using that $n \geq 4$, if $l \geq 2$ then the degree is negative. There could however be elements of degree 0 for $l \in \{0, 1\}$. Uf we go through the degree-counting argument carefully, we see that one-loop graphs of highest degree must have decorations of degree 1. This cannot happen because $R^1 = 0$.

We therefore deduce that W is homotopic to W_0 (relative to w_0) under the assumption that $\dim M \geq 4$ and that M is simply connected (by choosing a model such as $R^1 = 0$). □

Definition 4.45 We define a new graph complex mGraphs_R^0, exactly as aGraphs_R but replacing w by w_0 and W by W_0.

Corollary 4.46 *The two comodules in symmetric collections of CDGAs* mGraphs_R *and* mGraphs_R^0 *are quasi-isomorphic.*

This allows us to build a model mGraphs_R^0 that depends only on the real homotopy type of M, under dimension and connectivity assumptions.

4.3.5 Quasi-Isomorphism

In this section, we show the following:

Proposition 4.47 *The morphisms from* $\text{aGraphs}_{R_\partial}(U)$ *to* $\Omega_{\text{PA}}^*(\text{aFM}_N(U))$ *and from* $\text{mGraphs}_R(U)$ *to* $\Omega_{\text{PA}}^*(\text{mFM}_M(U))$ *defined using our integration procedures are quasi-isomorphisms.*

Proof The proofs are similar to the one in Sect. 3.5. Let us deal with $\text{aGraphs}_{R_\partial}$, as the case of mGraphs_R is almost identical. We work by induction. The result is clear for $\#U = 0$ and the case $\#U = 1$ is proved by an explicit homotopy (see the homotopy h in the proof of Lemma 3.123). For the induction step, assume that aGraphs$_{R_\partial}(U) \to \Omega^*_{PA}(aFM_N)$ is a quasi-isomorphism and let us assume $U_+ = U \sqcup \{u\}$. Consider the diagram (where the bottom line is the homotopy cofiber):

$$
\begin{array}{ccc}
\text{aGraphs}_{R_\partial}(U) & \xrightarrow{\ \sim\ } & \Omega^*_{PA}(aFM_N(U)) \\
\downarrow & & \downarrow \\
\text{aGraphs}_{R_\partial}(U_+) & \longrightarrow & \Omega^*_{PA}(aFM_N(U_+)) \\
\downarrow & & \downarrow \\
\text{aGraphs}_{R_\partial}(U_+) \otimes^L_{\text{aGraphs}_{R_\partial}(U)} \mathbb{R} & \longrightarrow & \Omega^*_{PA}(aFM_N(U_+)) \otimes^L_{\Omega^*_{PA}(aFM_N(U))} \mathbb{R}
\end{array}
$$

$$(4.34)$$

By the five lemma, it is enough to show that the top and bottom lines are quasi-isomorphisms to show that the middle line is. For the top line, this is precisely the induction hypothesis, so let us focus on the bottom line. The CDGA on the RHS is a model for the fiber of the fiber bundle $aFM_N(U_+) \to aFM_N(U)$ thanks to classical results in rational homotopy theory (see [FHT01, Theorem 15.3]). The fiber is homeomorphic to $(N \times \mathbb{R}) \setminus U$, i.e., the cylinder on N with $\#U$ points removed. The cohomology of this fiber is simply $H^*(N) \oplus \mathbb{R}[1 - n]^{\oplus U}$.

Moreover, since $aGraphs_{R_\partial}(U_+)$ is a quasi-free $aGraphs_{R_\partial}(U)$-module, the derived tensor product in the LHS is quasi-isomorphic to the usual tensor product. This tensor product can be identified to graphs whose internally connected components touch $u \in U_+$. This complex can be filtered by the number of edges incident to the vertex u, denoting the several components (with obvious notation) $V_0 \oplus V_1 \oplus V_{\geq 1}$. We have of course $V_0 = \mathbb{R}$. On the first page of the spectral sequence, the differential $d : V_1 \to V_{\geq 1}$ is surjective. Its kernel consists in the graphs whose vertex u is either decorated by $R_\partial^{>0}$, or of valence 1 and is connected to another external vertex. The cohomology of $N \times \mathbb{R} \setminus U$ can be found on the next page, and since the map in the forms is surjective in cohomology, none of these classes can disappear, which allows us to conclude that the next morphism in the induction step is also a quasi-isomorphism. $\qquad\square$

This allows us to establish the following theorem:

Theorem 4.48 ([CILW18]) *Let M be a simply connected compact SA manifold with boundary, of dimension at least 4. Let $(\lambda : B \to B_\partial)$ be a Poincaré–Lefschetz duality model of the pair $(M, \partial M)$. Then a model for the monoid and its module (mFM_M, aFM_N) is given by $(mGraphs^0_B, aGraphs^0_{B_\partial})$.*

We also have the following homotopy invariance results:

Corollary 4.49 *If N is a closed oriented SA manifold, then the real homotopy type of $\text{Conf}_{N \times \mathbb{R}}(r)$ only depends on the real homotopy type of N, for all $r \geq 0$.*

Corollary 4.50 *If M is a simply connected compact SA manifold with boundary of dimension* $\dim M \geq 4$, *then the real homotopy type of* $\mathrm{Conf}_M(r)$ *only depends on the real homotopy type of* $(M, \partial M)$, *for all* $r \geq 0$.

Remark 4.51 As Example 2.23 shows, it is not reasonable to expect the homotopy type of $\mathrm{Conf}_M(r)$ to only depend on the homotopy type of M for open manifolds.

4.4 Perturbed Lambrechts–Stanley Model

In this section, we will define an analogue of the Lambrechts–Stanley model for configuration spaces of compact manifolds with boundary. We first start by defining a naive version as a direct adaptation of the Lambrechts–Stanley model, and we show that it has the correct Betti numbers. Then we define a perturbed version of that naive adaptation, and we show that it is, in fact, a real model for the configuration spaces.

Let $(M, \partial M)$ be a compact manifold with boundary, of dimension $\dim M = n$. Let $(B \xrightarrow{\lambda} B_\partial)$ be a Poincaré–Lefschetz duality model of $(M, \partial M)$, and write $K = \ker \lambda$ and $P = B/I$ (with the notation of Sect. 4.2). We recall that we have a diagonal class $\Delta_P \in P \otimes P$ which is the image of the class Δ_{KP} induced by the duality between K and P.

4.4.1 Computation of the Homology

We can first generalize the definition of the Lambrechts–Stanley model in an obvious way.

Definition 4.52 The *Lambrechts–Stanley CDGA* associated with $B \xrightarrow{\lambda} B_\partial$ is:

$$\mathsf{G}_P(U) := \left(P^U \otimes (S(\omega_{ij})_{i \neq j \in U} / I, d \right) \tag{4.35}$$

with $d\omega_{ij} = p_{ij}^*(\Delta_P)$ and $I = \left(\omega_{ij}^2, \omega_{ji} - (-1)^n \omega_{ij}, \omega_{ij}\omega_{jk} + \omega_{jk}\omega_{ki} + \omega_{ij} \right)$.

The goal of this section is to prove the following theorem.

Theorem 4.53 ([CILW18]) *The cohomology of* $\mathsf{G}_P(U)$ *is isomorphic to the cohomology of* $\mathrm{Conf}_M(U)$ *as a graded vector space.*

Example 4.54 Let $M = \mathbb{D}^n$ and $\partial M = \mathbb{S}^{n-1}$. A Poincaré–Lefschetz duality model of $(M, \partial M)$ is given in Example 4.10. In particular, we have $P = \mathbb{R}$ and $\Delta_P = 0$. We then find $\mathsf{G}_P(U) = H^*(\mathrm{Conf}_{\mathbb{R}^n}(U))$. This is consistent with the fact that $\mathrm{Conf}_{\mathbb{D}^n}(U) \simeq \mathrm{Conf}_{\mathbb{R}^n}(U)$.

Example 4.55 This construction can also be applied to a closed manifold Z with a point removed, $M = Z \setminus *$. Let A be a Poincaré duality model of W. We saw in Example 4.11 that a PLD model for $(M, \partial M)$ is given by $B = A \oplus \mathbb{R}\langle v_{n-1} \rangle$, $B_\partial = \mathbb{R} \oplus \mathbb{R}\langle v \rangle$. The kernel K is the augmentation ideal $\bar{A} = \ker(A \to \mathbb{R})$, and the quotient P is $A/(\text{vol}_A)$. The diagonal class Δ_P is just the image of Δ_A in the quotient. We thus find, on the one hand, a small CDGA $G_P(U)$ whose cohomology is the cohomology of $\text{Conf}_M(U)$.

On the other hand, we know, using the Fadell–Neuwirth fibration (see Remark 2.91), that $\text{Conf}_M(U)$ is the fiber of the projection $p_* : \text{Conf}_W(U_+) \to W$, where $U_+ = U \sqcup \{*\}$. We know that a model of $\text{Conf}_W(U_+)$ is $G_A(U_+)$, and it is clear that a model of p_* is the inclusion $A \to G_A(U_+)$ in position $*$. By general considerations about Sullivan models (see e.g., [FOT08, Theorem 2.64]), a model for the fiber of p_* is thus given by the tensor product $G_A(U_+) \otimes_A \mathbb{R}$. Note that in this tensor product, $d\omega_{*i} = p_i^*(\text{vol}_A)$. This is consistent with the previous result, as we thus have an obvious quasi-isomorphism $G_A(U_+) \otimes_A \mathbb{R} \to G_P(U)$.

The proof of Theorem 4.53 relies on two other theorems. On the one hand, we will use the methods of [LS08a], which allows us to compute the homology of a "configuration space"-type space, i.e., one that is obtained by removing subspaces indexed by pairs of integers and which intersect as diagonals do. On the other hand, we will use [CLS18, Theorem 1.1], which allows to compute the homology of the complement of a subspace in a manifold with boundary. More concretely, [CLS18, Theorem 1.1] tells us that if W is a compact manifold with boundary and that $X \subset W$ is a sub-polyhedron, then we can compute the rational homology of $W \setminus X$ if we know a rational model of the following square:

$$
\begin{array}{ccc}
\partial W & \lhook\joinrel\longrightarrow & W \\
\uparrow & & \uparrow \\
\partial_W X := X \cap \partial W & \lhook\joinrel\longrightarrow & X
\end{array}
\tag{4.36}
$$

In our case, $W = M^U$ and $X = \bigcup_{i \neq j \in U} \Delta_{ij}$, where $\Delta_{ij} = \{x \in M^U \mid x_i = x_j\}$. The methods used in [LS08a] allow to simplify the computation of a model of the previous square by expressing it as the "total cofiber". The idea is to write X as the colimit of a diagram indexed by graphs. To simplify notation, let us put $r = \#U$. Let $E = \{(i, j) \mid 1 \leq i < j \leq r\}$ and let $\mathcal{P}$ be the set of subsets of E, ordered by reverse inclusion. An element $\gamma \in \mathcal{P}$ can be seen as a graph with r vertices, with an edge between i and j if and only if $(i, j) \in \gamma$. We then have a functor $\nabla : \mathcal{P} \to \text{Top}$ given by

$$
\nabla(\gamma) := \bigcap_{(i,j) \in E_\gamma} \Delta_{ij}.
\tag{4.37}
$$

In particular, we have $\nabla(\varnothing) = M^r$ and

$$\gamma' \supset \gamma \implies \nabla(\gamma') \subset \nabla(\gamma). \tag{4.38}$$

The space $\nabla(\gamma)$ is homeomorphic to the product $M^{\pi_0(\gamma)}$, where γ is viewed as a graph and $\pi_0(\gamma)$ is its set of connected components. We have that $X = \bigcup_{1 \le i < j \le r} \Delta_{ij}$ is given by the following colimit (i.e., union):

$$X = \operatorname*{colim}_{\gamma \in \mathcal{P}} \nabla(\gamma) = \operatorname*{colim}_{\gamma \in \mathcal{P}} M^{\pi_0(\gamma)}. \tag{4.39}$$

The diagram $\gamma \mapsto M^{\pi_0(\gamma)}$ satisfies a technical condition that roughly speaking means that it is well-behaved with respect to colimits and homotopy:

Proposition 4.56 *The diagram $\gamma \mapsto \nabla(\gamma)$ is cofibrant in the Reedy model structure of diagrams.*

We refer to [Hov99, Section 5] for details on this notion, that we are not going to explain precisely here. Checking that the diagram is Reedy cofibrant involves the investigation of the latching spaces, defined, for $\gamma \in \mathcal{P}$, by:

$$L_\gamma \nabla := \operatorname*{colim}_{\gamma' \supset \gamma, \ \gamma' \ne \gamma} M^{\pi_0(\gamma)}. \tag{4.40}$$

There exists a canonical map $\iota_\gamma : L_\gamma \nabla \to \nabla(\gamma)$ that is merely the inclusion. In order to check that the diagram is Reedy cofibrant, it must be checked that ι_γ is a cofibration for every γ. But here, this is clear, as one can explicitly describe $L_\gamma \nabla$ by $L_\gamma \nabla = \nabla(\gamma)$ if γ is connected, and otherwise:

$$L_\gamma \nabla \cong \{x \in \nabla(\gamma) = M^{\pi_0(\gamma)} \mid \exists [i] \ne [j] \in \pi_0(\gamma) \text{ s.t. } x_i = x_j\}. \tag{4.41}$$

The fact that the diagram defining X is Reedy cofibrant has the corollary that X is actually the homotopy colimit of the diagram $\gamma \mapsto M^{\pi_0(\gamma)}$. It is a general fact that homotopy colimits of topological spaces are modeled by homotopy limits of CDGAs (which merely stems from the fact that $\Omega_{\mathrm{PL}}^* \dashv \langle - \rangle$ is a Quillen adjunction). We thus deduce that:

Corollary 4.57 *Let M be a compact manifold with boundary, (B, B_∂) a PLD model of $(M, \partial M)$, $r \ge 0$ an integer, and $X \subset M^r$ be the fat diagonal. A rational model for X is given by $\lim_{\gamma \in \mathcal{P}^{\mathrm{op}}} B^{\otimes \pi_0(\gamma)}$ with the notation as above.*

In order to apply the theorem of [CLS18], all that remains to do is to find a model of $\partial_{M^r} X = X \cap \partial M^r$.

Lemma 4.58 *With the same notation as in the previous lemma, we have:*

$$X \cap \partial M^r = \operatorname*{colim}_{\gamma \in \mathcal{P}} \operatorname*{colim}_{\varnothing \subsetneq S \subset \pi_0(\gamma)} (\partial M)^S \times M^{\pi_0(\gamma) \setminus S}. \tag{4.42}$$

Proof This follows from the fact that the boundary of ∂M^r is characterized as r-uples where at least one point is in the boundary. In the decomposition of X of Eq. (4.39), if $x \in \nabla(\gamma)$ is such that $x_i \in \partial M$, then $x_j = x_i \in \partial M$ for any j in the component of i in the graph γ. □

Lemma 4.59 *With the notation as in the previous lemmas, and for every integer i, the square on the left is a model for the square on the right:*

$$
\begin{array}{ccc}
B \xleftarrow{\mu_B^{(i)}} B^{\otimes i} \\
\downarrow \qquad \downarrow \\
B/K \xleftarrow{\mu_B^{(i)}} B^{\otimes i}/K^{\otimes i}
\end{array}
\quad \text{is a model of} \quad
\begin{array}{ccc}
M \xhookrightarrow{\delta} M^i \\
\uparrow \qquad \uparrow \\
\partial M \xhookrightarrow{\delta} \partial(M^i)
\end{array}.
$$

$$\text{(4.43)}$$

Proof The proof is by induction on i. It is obvious for $i = 1$, because $B_\partial = B/K$ and we have assumed that $\lambda : B \to B_\partial$ is a model of $\partial M \hookrightarrow M$. For the induction step from i to $i + 1$, we consider the following diagram:

$$
\begin{array}{ccc}
M \times M^i & \xleftarrow{\hspace{3cm}} & M \\
\uparrow & & \uparrow \\
\partial(M \times M^i) \xleftarrow{} M \times \partial(M^i) & & \\
\uparrow \qquad\qquad\qquad \uparrow & & \\
(\partial M) \times M^i \xleftarrow{} (\partial M) \times (\partial(M^i)) \xleftarrow{} \partial M
\end{array}
$$

$$\text{(4.44)}$$

The square in the bottom left is a pushout, i.e., a point belongs to the boundary of the product if and only if at least one of its factors belongs to the boundary. Since the inclusion $\partial M \hookrightarrow M$ is a cofibration, the square is actually a homotopy pushout. Let us denote by P the pullback:

$$
\begin{array}{ccc}
P & \dashrightarrow & B \otimes (B^{\otimes i}/K^{\otimes i}) \\
\vdots & \lrcorner & \downarrow \\
B_\partial \otimes B^{\otimes i} & \longrightarrow & B_\partial \otimes (B^{\otimes i}/K^{\otimes i})
\end{array}
$$

$$\text{(4.45)}$$

Since models of homotopy colimits (which include homotopy pushouts) are limits (which include pullbacks), we find that a model of the diagram of Eq. (4.44) is given by:

$$
\begin{array}{ccc}
B \otimes B^{\otimes i} & \xrightarrow{\hspace{4cm}} & B \\
\downarrow & & \downarrow \\
P \longrightarrow B \otimes (B^{\otimes i}/K^{\otimes i}) & & \\
\downarrow \qquad\qquad\qquad \downarrow & & \downarrow \\
B_\partial \otimes B^{\otimes i} \longrightarrow B_\partial \otimes (B^{\otimes i}/K^{\otimes i}) \longrightarrow B_\partial
\end{array}
$$

$$\text{(4.46)}$$

Concretely, we have that:

$$P = \left\{ (x, y) \in (B_\partial \otimes B^{\otimes i}) \times (B \otimes (B^{\otimes i}/K^{\otimes i})) \mid (\mathrm{id} \otimes \pi_i)(x) = (\lambda \otimes \mathrm{id})(y) \right\}, \tag{4.47}$$

where $\pi_i : B^{\otimes i} \to B^{\otimes i}/K^{\otimes i}$ is the quotient map. It is not hard to see that the natural map $B^{\otimes(i+1)} \to P$ is surjective, with kernel $K^{\otimes(i+1)}$. It follows that $P \cong B^{\otimes(i+1)}/K^{\otimes(i+1)}$ and we obtain the result. □

Corollary 4.60 *With the same notation as above, a model for the square of the Eq. (4.36) where $W = M^r$ and X is the fat diagonal is given by:*

$$
\begin{array}{ccc}
B^{\otimes k} & \xrightarrow{\ \ \alpha_k\ \ } & B^{\otimes k}/K^{\otimes k} \\
\downarrow{\scriptstyle \xi_k} & & \downarrow \\
\lim_{\gamma \in \mathcal{P}^{\mathrm{op}}} B^{\otimes \pi_0(\gamma)} & \xrightarrow{\ \ \beta_k\ \ } & \lim_{\gamma \in \mathcal{P}^{\mathrm{op}}} B^{\otimes \pi_0(\gamma)}/K^{\otimes \pi_0(\gamma)}
\end{array} \tag{4.48}
$$

Proof (Sketch of Proof of Theorem 4.53) Applying the main theorem of [CLS18], we find that the cohomology (as graded vector space) of $W \setminus X = \mathsf{Conf}_M(r)$ is calculated by the homology of the cone of the induced map:

$$\mathrm{cone}\big((\mathrm{hoker}\,\beta_r)^\vee[-nr] \xrightarrow{\ \bar{\xi}_r\ } (K^{\otimes r})^\vee[-nr]\big), \tag{4.49}$$

where the homotopy kernel was defined in Definition 4.3. Using the duality between K and P, the space $(K^{\otimes r})^\vee[-nr]$ is isomorphic to $P^{\otimes r}$, which is a model for M^r. Moreover, a careful application of Poincaré–Lefschetz duality shows that $(\mathrm{hoker}\,\beta_r)^\vee[-nr]$ is a model for $\bigcup \Delta_{ij}$. The cone thus has the homology of the complement, which is $\mathsf{Conf}_M(r)$. □

4.4.2 Perturbed Model

Unfortunately, the Lambrechts–Stanley CDGA $\mathsf{G}_P(U)$ is generally not a model of $\mathsf{Conf}_M(U)$. In this section, we explain how to fix this issue. The following example gives an idea of why (although it does not satisfy our connectedness assumptions).

Example 4.61 Let $M = \mathbb{S}^1 \times [0, 1]$ be the cylinder with boundary $\mathbb{S}^1 \times \{0, 1\}$. The pair $(M, \partial M)$ admits a Poincaré–Lefschetz duality model with the following data:

- $B_{\partial M} = H^*(\partial M)$ is spanned by $1, t, d\varphi, t\,d\varphi$, with $t^2 = t$;
- $B = \langle 1, t, dt, d\varphi, t, dt, d\varphi, dt \wedge d\varphi \rangle$, so in particular $2t\,dt = dt$.
- $P = H^*(M) = \langle 1, d\varphi \rangle$;
- $K = H^*(M, \partial M) = \langle dt, dt \wedge d\varphi \rangle$.

Then the symmetry relations in $G_P(2)$ imply that $(d\varphi \otimes 1)\omega_{12} = (1 \otimes \varphi)\omega_{12}$. However, this relation is intuitively not correct. Indeed, M is a plane from which we have removed a point, so we have a Fadell–Neuwirth fibration:

$$\mathsf{Conf}_M(2) \longhookrightarrow \mathsf{Conf}_{\mathbb{R}^2}(3) \longrightarrow\!\!\!\!\rightarrow \mathbb{R}^2 \tag{4.50}$$

from which one deduces $\mathsf{Conf}_M(2) \simeq \mathsf{Conf}_{\mathbb{R}^2}(3)$. The class $d\varphi \otimes 1$ corresponds to ω_{13} and $1 \otimes d\varphi$ corresponds to ω_{23}. The previous relation is therefore not the usual Arnold relation: we have to add the term $d\varphi \otimes d\varphi$ to correct it.

Fortunately, there is a way to modify $G_P(r)$ to overcome this problem. We consider the quotient of $\mathsf{mGraphs}^0_A(r)$ by the dg-ideal generated by graphs containing at least one internal vertex. More concretely, consider the section $\sigma \in B \otimes B_\partial$ (see Definition 4.16), which is projected onto an element $\sigma_P \in P \otimes B_\partial$. Let us write $\sigma_P = \sum_i \sigma'_i \otimes \sigma''_i$ in terms of elementary tensors.

Definition 4.62 The *perturbed Lambrechts–Stanley model* associated to the previous data is:

$$\tilde{G}_P(U) := \left(P^{\otimes U} \otimes S(\tilde{\omega}_{ij})_{i,j\in u}/\tilde{I}, d\right) \tag{4.51}$$

where the ideal $\tilde{I}$ is generated by the relations $\tilde{\omega}^2_{ij} = \tilde{\omega}_{ii} = 0$, and, for every $T \subset U$ of cardinal 2 or 3:

$$\sum_{v\in T} \pm\left(\iota_v(\pi(b)) \cdot \prod_{v\neq v'\in T} \tilde{\omega}_{vv'}\right) + \sum_{i_1,\dots,i_k} \pm\varepsilon_\partial\left(\rho(b)\prod_{v\in T}\sigma''_{i_v}\right)\prod_{v\in T}\iota_v(\sigma'_{i_v}). \tag{4.52}$$

The last relations are simply the images of the differential in $\mathsf{aGraphs}_B(U)$ of graphs with a single internal vertex. Let us now make these relations more explicit. For $\#T = 2$, this relation becomes, for $b \in B$:

$$(b \otimes 1)\tilde{\omega}_{12} - (-1)^n(1 \otimes b)\tilde{\omega}_{21} + \sum_{i,j} \pm\varepsilon_\partial(\sigma''_i\sigma''_j)\sigma'_i \otimes \sigma'_j = 0 \in \tilde{G}_P(2). \tag{4.53}$$

In particular, for $b = 1$, we see that $\tilde{\omega}$ is symmetric up to terms of weight 0, and that similarly the symmetry relation holds up to terms of weight 0 (where by weight we mean the degree of a monomial in $S(\tilde{\omega}_{ij})$). For $\#T = 3$ and $b = 1$, we get a perturbation of Arnold's relation, i.e., the Arnold relation up to terms of weight ≤ 2. For $\#T = 3$ and a generic element $b \in B$, we get a mix of the perturbed Arnold and symmetry relations.

Lemma 4.63 *There is an isomorphism of dg-modules $\tilde{G}_P(U) \cong G_P(U)$. In particular, the cohomologies $H^*(\tilde{G}_P(U))$ and $H^*(\mathsf{Conf}_M(U))$ are isomorphic as graded vector spaces.*

Proof The standard basis of $H^*(\mathsf{Conf}_{\mathbb{R}^n}(r))$ is given by words of the type $\omega_{i_1 j_1} \ldots \omega_{i_k j_k}$, with $1 \leq i_1 < \cdots < i_k \leq r$ and $i_l < j_l$ for all l (see Lemma 2.89). By choosing a basis of P, we thus get a basis of $\mathsf{G}_P(r)$. One checks without difficulty that by replacing ω with $\tilde{\omega}$ in that basis, we obtain a basis of $\tilde{\mathsf{G}}_P(r)$. Moreover, the map $\mathsf{G}_P(r) \to \tilde{\mathsf{G}}_P(r)$ thus obtained clearly preserves the differential.

□

Remark 4.64 In many cases, we have $\mathsf{G}_P = \tilde{\mathsf{G}}_P$. It is for example the case if $M = N \setminus \mathbb{D}^n$ where N is a closed manifold. Indeed, in this case, $\sigma_P = 1 \otimes v$ and $v^2 = 0$ so all the extra terms with a vanish in the perturbed relations.

Proposition 4.65 *The quotient map* $\mathsf{mGraphs}^0_B(r) \to \tilde{\mathsf{G}}_P(r)$ *is a quasi-isomorphism of CDGA.*

Proof The proof is almost the same as the proof of Proposition 3.118 in the case without a boundary. Indeed, as soon as we filter by $\#E - \#V$, the perturbed relations coincide with the usual relations.

□

We thus get the following theorem immediately, thanks to Theorem 4.48.

Theorem 4.66 ([CILW18]) *Let M be a smooth compact simply connected manifold of dimension at least 7. Suppose that $(M, \partial M)$ admits a PLD model (B, B_∂) and let $P = B/I$ be the quotient as defined in Sect. 4.2. Then $\tilde{\mathsf{G}}_P(r)$ is a real model of $\mathsf{Conf}_M(r)$.*

Remark 4.67 In dimension ≤ 6, if the manifold and its boundary are both simply connected, then both are formal. (Note however that the inclusion $\partial M \subset M$ may not be formal.) We can then show in the same way that $\tilde{\mathsf{G}}_{H^*(M)}(r)$ is a model of $\mathsf{Conf}_M(r)$.

Chapter 5
Configuration Spaces and Operads

In this final chapter, we explain some of the connections that exist between configuration spaces and operads. Briefly, an operad is a device that encodes a category of algebras, such as associative algebras, commutative algebras, Lie algebras, and so on. In topology, they were introduced in the study of iterated loop spaces (see Sect. 5.2), which have a structure encoded by a certain class of operad, the little disks operads. These operads are central to the theory and appear in many applications (see the discussion after Theorem 5.46). They are closely related to configuration spaces of Euclidean spaces, as we will explain in Sect. 5.4. More precisely, the Fulton–MacPherson compactifications of these configuration spaces, which we introduced in Sect. 3.2, assemble to form an operad which is weakly homotopy equivalent to the little disks operad. Moreover, if a manifold M is framed, then there exists an action of the Fulton–MacPherson operad on the collection of Fulton–MacPherson compactification of the configuration spaces of M. We show in Sect. 5.4 that the models of configuration spaces, which we obtained in Chap. 3, are compatible with the operadic action. This action is important in several applications, and in particular factorization homology, which we introduce in Sect. 5.1. We end the chapter in Sect. 5.5 with an example of computation of factorization homology using the results previously obtained.

5.1 Motivation: Factorization Homology

The homotopy types of configuration spaces can be used to define homeomorphism invariants of manifolds. Indeed, it is clear that if two manifolds are homeomorphic, then their configuration spaces are also homeomorphic and therefore have the same homotopy type. For a manifold M, one can therefore study the homotopy invariants (homology, homotopy groups, etc) of $\mathrm{Conf}_M(r)$ to produce invariants that are generally finer than homotopy invariants of M. These invariants are

© The Author(s), under exclusive license to Springer Nature Switzerland AG 2022
N. Idrissi, *Real Homotopy of Configuration Spaces*, Lecture Notes
in Mathematics 2303, https://doi.org/10.1007/978-3-031-04428-1_5

moreover functorial with respect to manifold embeddings, as an embedding of manifolds $M \hookrightarrow N$ produces maps $\mathsf{Conf}_M(r) \to \mathsf{Conf}_N(r)$ that induce themselves morphisms between homotopy invariants.

The objective of this section is to explain how to produce another kind of manifold invariants from configuration spaces. This new invariant, called "factorization homology" of the manifold, is motivated by physical considerations. More precisely, its idea comes from topological quantum field theory.

Informally, the idea behind factorization homology is to add decorations to the points in a configuration. One can for example think of the electric charge: in this case, the decorations are numerical invariants, that represents the electric charge of a particle. Calculating the factorization homology on M with coefficients in some space of decorations A consists in averaging over all possible positions of the points, with the rule that if points collide in a configuration, then their decorations (e.g., the electrical charge) are added. If we only have numerical decorations for our points, then we get invariants related to higher Hochschild homology [Pir00], which is in general a coarse invariant. However, the space of decorations A can be more complicated, and in particular the operation used to merge decorations (when two particles collide) can be non-commutative.

An important question is thus the structure of the coefficients: how to add the decorations of points that meet each other? In dimension 1, there are only two ways for a pair of points to meet (up to homotopy): from left to right or from right to left. In higher dimension, though, things become much more complicated. Two given points can meet in an infinite number of different ways, even up to homotopy. For example, in dimension 2, all that matters (up to homotopy) is the angle with which the points meet, so we get a whole circle worth of possibilities of collisions. More generally, in dimension $n \in \mathbb{N}$, the possibilities of collisions take the shape of $\mathbb{S}^{n-1}$. All these ways of colliding are homotopic (because $\mathbb{S}^{n-1}$ is connected for $n \geq 2$), but there can be several essentially different homotopies between two ways of colliding. For example with $n = 2$, two points can meet from left to right, or from right to left on the horizontal axis. These two ways of colliding are homotopic through two different homotopies: we can rotate the configuration clockwise, or we can rotate it counter-clockwise. In higher dimensions still, we get even more subtle behavior: in e.g., dimension 3, all homotopies between two different ways of colliding are homotopic, but the homotopies between the homotopies themselves can differ (summarized by the fact that $\pi_2(\mathbb{S}^2) \neq 0$).

The picture becomes even more complicated when more than two points collide at once. The axiomatization of the relations that the different multiplications must satisfy is encoded by the theory of operads. We will briefly introduce this theory in Sect. 5.2. We will then see how to give a precise definition of factorization homology using Fulton–MacPherson compactifications.

Factorization homology has properties reminiscent of the Eilenberg–Steenrod axioms of usual homology, which explains the name. Let us denote by $\int_M A$

the factorization homology of M with coefficients in A. Factorization homology satisfies the following axioms:

1. The homology of $\mathbb{R}^n$ is simply the space of coefficients, i.e., $\int_{\mathbb{R}^n} A \simeq A$. Indeed, since $\mathbb{R}^n$ is contractible, we can bring all the points of a configuration back to the origin, adding their decorations as we go along. This mirrors the classical axiom that $H^*(\text{pt}) = R$ where R is the ring of coefficients.

2. If $M = M' \sqcup M''$ is the disjoint union of two manifolds, then the homology of M is the tensor product of the homologies of M' and M'', i.e., $\int_{M' \sqcup M''} A \simeq \int_{M'} A \otimes \int_{M''} A$. This mirrors $H^*(U \sqcup V) \cong H^*(U) \oplus H^*(V)$ (and also explains why factorization homology can be seen as a non-linear version of homology, as the tensor product replaces direct sum).

3. Finally, factorization homology satisfies an axiom of the Mayer–Vietoris kind. If $M = M' \cup_{N \times \mathbb{R}} M''$ is obtained by gluing two manifolds with common boundary N along a collar around their boundary, then one gets a formula that expresses $\int_M A$ in terms of $\int_{M'} A$, $\int_{M''} A$, and $\int_{N \times \mathbb{R}} A$. This essentially follows from the formula which allows the calculation of the configuration spaces of a manifold obtained by joining two manifolds along their boundary, see Eq. (4.3).

In fact, thanks to a theorem of Francis [Fra13], factorization homology is (up to homotopy) the unique functor that satisfies these axioms.

5.1.1 Historical Remarks

Factorization homology has been studied by many authors, in different frameworks and under different names. It was introduced under this name by Francis [Fra13] and developed in particular by Ayala and Francis [AF15] and Ayala et al. [AFT17]. It is notably inspired by the topological chiral homology of Lurie [Lur09a; Lur09b], which is itself a homotopical analogue of the chiral homology of Beilinson and Drinfeld [BD04] and influenced by the work of Segal [Seg73; Seg04]. It is also linked to the blob homology of Morrison and Walker [MW12]. The idea of factorization homology is close to that of the decorated configuration spaces of Salvatore [Sal01], who has also proved the link with the compactifications of Fulton–MacPherson. Finally, factorization homology is strongly related to factorization algebras, which are a kind of cosheaf version of factorization homology; see in particular Costello and Gwilliam [CG17a; CG17b]. One may refer to Ginot [Gin15] for a broad overview.

Remark 5.1 In Chap. 2, we mentioned another application of configuration spaces: Goodwillie–Weiss embedding calculus (see Example 2.8). This calculus is in some sense "adjoint" to factorization homology. While factorization homology is encoded by a derived tensor product (see Definition 5.99 below), theorems of Boavida de Brito and Weiss [BW18] and Turchin [Tur13] imply that the space of embeddings $\text{Emb}(M, N)$ can be expressed as a derived mapping space.

5.2   Introduction to Operads

In this section, we will briefly introduce the theory of operads. One may refer to Loday and Vallette [LV12] and Fresse [Fre17a, Part I(a)] for more complete references.

An operad is an object that governs a category of "algebras" in a wide sense, e.g., associative algebras, commutative algebras, or Lie algebras. If we work by analogy with group theory, an operad is to its category of algebras what a group is to its category of representations. Historically, categories of algebras have often been defined by generators and relations. For example, an associative algebra is the data of a vector space equipped with a binary product which must satisfy the relation of associativity. The idea of the theory of operads is that, just like there is a group that can be studied on its own when we are given a category of representations, there is an operad that can be studied on its own when we are given a category of algebras, without necessarily being concerned with a fixed presentation. This allows the introduction of numerous notions whose interest in group theory is not to be demonstrated: morphisms, sub-objects, quotients, extensions, etc. Just like for groups, these notions inform us about the categories of algebras associated to the operads that we study.

The theory of operads goes back a long way. An operad is a special case of PROP, a notion introduced by Mac Lane [Mac65] in the 1960s, and they are also a special case of the analyzers of Lazard [Laz55]. The notion of operad itself was initially introduced in algebraic topology to study iterated loop spaces at the end of the 1960s by Boardman and Vogt [BV68, BV73] and May [May72]. The term "operad" itself was introduced by May [May72] and is a contraction of the words *operation monad*. The operads that appeared in [BV68; May72] are the little disks operads D_n that we are going to study in Sect. 5.3.1. The components of the operad of little intervals D_1 (see below) are homotopy equivalent to the associahedra of Stasheff [Sta61], and the operadic structure of associahedra already appears implicitly in his work. After this initial introduction, and particularly since the mid-1990s, interest in operads has grown considerably. This stems from the discovery made by Ginzburg and Kapranov [GK94], following ideas of Kontsevich [Kon93], who have shown that some phenomena of duality in algebra had an operadic interpretation via Koszul duality. Since then, many applications of operads in several fields of mathematics have been discovered.

5.2.1   Definition of Operads

The algebraic structures encoded by operads are of a very specific type. They are structures that can be described using operations with several inputs and exactly one output, and the relations between these operations cannot involve repeated input variables. Let us start by illustrating operads with a fundamental example.

Remark 5.2 What we are going to define are single-colored symmetric operads. There are many other variants of the notion of operad, see e.g., [LV12, Section 13.14] for an overview.

Remark 5.3 Most of what follows can be written within any symmetric monoidal category $(\mathsf{C}, \otimes, \mathbb{I})$. Typical examples include sets, topological spaces, vector spaces, (co)chain complexes, or CDGAs. In order to have an easier time writing down formulas below, we will assume that C belongs to this list of examples. This allows us to write pointwise formulas, rather than explicitly writing down associators and braidings.

In what follows, we are going to write down a permutation $\sigma \in \mathfrak{S}_k$ of the kth symmetric group as a list of numbers, $(\sigma(1), \ldots, \sigma(k))$.

The prototype of an operad is the *endomorphism operad* End_X of an object $X \in \mathsf{C}$. This operad is given by a collection of operations, graded by the number of inputs. In more detail, we set:

$$\mathrm{End}_X = \{\mathrm{End}_X(k)\}_{k \geq 0} := \{\mathrm{Hom}(X^{\otimes k}, X)\}_{k \geq 0}. \tag{5.1}$$

This collection is equipped with the following structure, which stems from composition of morphisms and will form the template for defining operads.

- The collection has a unit $\mathrm{id}_X \in \mathrm{End}_X(1)$.
- For each integer $k \geq 0$, the symmetric group $\mathfrak{S}_k$ acts on $\mathrm{End}_X(k)$ by permuting the inputs. If $f \in \mathrm{End}_X(k)$ and $\sigma \in \mathfrak{S}_k$, then we can define $\sigma \cdot f \in \mathrm{End}_X(k)$ by:

$$(f \cdot \sigma)(x_1, \ldots, x_k) := f(x_{\sigma^{-1}(1)}, \ldots, x_{\sigma^{-1}(k)}). \tag{5.2}$$

- For all integers $0 \leq i \leq k$ and $l \geq 0$, one can compose the operations using the map $\circ_i : \mathrm{End}_X(k) \otimes \mathrm{End}_X(l) \to \mathrm{End}_X(k + l - 1)$ defined, for $f \in \mathrm{End}_X(k)$ and $g \in \mathrm{End}_X(l)$, by:

$$(f \circ_i g)(x_1, \ldots, x_{k+l-1}) := f(x_1, \ldots, x_{i-1}, g(x_i, \ldots, x_{i+l-1}), x_{i+1}, \ldots, x_{k+l-1}). \tag{5.3}$$

(If we are in an algebraic category such as chain complexes, then one must add a sign to this formula to account for the fact that g "passed over" the variables $x_1, \ldots, x_{i-1}$.) It is often useful to represent this operation graphically, using rooted trees as in the following picture:

$$\tag{5.4}$$

These operations are part of the structure of an operad and satisfy a certain number of relations, see Definition 5.5. These relations all arise from the sequential and parallel composition axioms, unitality, and equivariance of the composition of functions. They imply that any rooted tree whose vertices are decorated by elements of End_X (a vertex with k children being decorated by an element $\text{End}_X(k)$) and whose leaves are decorated by a permutation of integers defines a new element of End_X. For example, the following tree (where $\text{id}_X : X \to X$ is the identity) corresponds to the operation $(x_1, \ldots, x_5) \mapsto f(x_2, g(x_1, x_3, x_4), x_5)$:

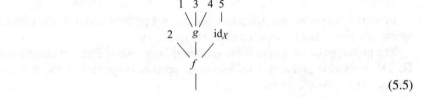

$$(5.5)$$

Based on these preliminaries, let us now define operads using the endomorphism operad as a template.

Definition 5.4 A *symmetric sequence* in C is an $\mathbb{N}$-indexed collection of objects $\mathsf{P} = \{\mathsf{P}(k)\}_{k \geq 0}$ of C equipped, for all $k \geq 0$, with an action of the symmetric group $\mathfrak{S}_k$ on $\mathsf{P}(k)$.

Definition 5.5 An *operad* in C is a symmetric sequence $\mathsf{P} = \{\mathsf{P}(k)\}_{k \geq 0}$ equipped with:

- a unit $\eta : \mathbb{I} \to \mathsf{P}(k)$, where $\mathbb{I}$ is the monoidal unit (whose image in pointwise notation will be denoted by $\text{id}_\mathsf{P} = \eta(1)$);
- a set of composition operations, for all $k \geq 0$ and $1 \leq i \leq k$:

$$\circ_i : \mathsf{P}(k) \otimes \mathsf{P}(l) \to \mathsf{P}(k + l - 1). \tag{5.6}$$

These operations should satisfy the following properties:

(Equivariance)   For all permutations $\sigma \in \mathfrak{S}_k, \tau \in \mathfrak{S}_l$,

$$(p \cdot \sigma) \circ_k (q \cdot \tau) = (p \circ_{\sigma(k)} q) \cdot (\sigma \circ'_k \tau), \tag{5.7}$$

where $\sigma \circ'_k \tau$ is the partial composition of the permutations σ and τ of Example 5.7.

(Unitality)   For any operation $p \in \mathsf{P}(k)$ and any index $1 \leq i \leq k$,

$$p \circ_i \text{id}_\mathsf{P} = p = \text{id}_\mathsf{P} \circ_1 p. \tag{5.8}$$

(Sequential composition)   for any operations $p \in P(k)$, $q \in P(l)$, and $r \in P(m)$, and for any indices $1 \leq i \leq k$, $1 \leq j \leq l$,

$$(p \circ_i q) \circ_{i+j-1} r = p \circ_i (q \circ_j r), \tag{5.9}$$

(Parallel composition)   for any operations $p \in P(k)$, $q \in P(l)$, and $r \in P(m)$, and for any indices $1 \leq i < j \leq k$,

$$(p \circ_i q) \circ_{j+l-1} r = (p \circ_j r) \circ_i q, \tag{5.10}$$

A morphism of operads $f : P \to Q$ is defined as a collection of σ-equivariant morphisms $\{f_k : P(k) \to Q(k)\}_{k \geq 0}$ that commute with the structure maps, i.e., $f_1(\mathrm{id}_P) = \mathrm{id}_Q$ and $f_{k+l-1}(p \circ_i p') = f_k(p) \circ_i f_l(p')$.

The following example is obvious, as the definition of operads was custom-tailored for the endomorphism operad:

Example 5.6 The endomorphism operad End_X of an object X is an operad.

Let us now describe classical examples of operads.

Example 5.7 We define an operad in sets denoted As by setting $\mathsf{As}(k) = \mathfrak{S}_k$ for all $k \geq 0$. The action of the symmetric group is simply group multiplication. The unit is the trivial permutation $(1) \in \mathfrak{S}_1$. The composition is defined as follows. For $\sigma = (\sigma_1, \ldots, \sigma_k) \in \mathfrak{S}_k$, $\tau = (\tau_1, \ldots, \tau_l) \in \mathfrak{S}_l$ and $1 \leq i \leq k$, we set:

$$\sigma \circ_i' \tau := \big(\sigma(1), \ldots, \sigma(i-1), \tau(1), \ldots, \tau(l), \sigma(i+1), \ldots, \sigma(k)\big) \in \mathfrak{S}_{k+l-1}. \tag{5.11}$$

and

$$\sigma \circ_{\sigma(i)} \tau := (\sigma^{-1} \circ_i' \tau^{-1})^{-1} \tag{5.12}$$

Example 5.8 We can define an operad Com by setting $\mathsf{Com}(k)$ to be a singleton for all $k \geq 0$. The action of the symmetric group is trivial. The unit $\eta = \mathrm{id}$ is the unique element of $\mathsf{Com}(1)$. Finally, the composition is the unique bijection $\mathsf{Com}(k) \times \mathsf{Com}(l) \cong \mathsf{Com}(k+l-1)$.

Remark 5.9 Definition 5.5 is the definition of an operad in terms of partial compositions. Operads can also be defined using total compositions (with adapted axioms):

$$\gamma : P(k) \otimes P(r_1) \otimes \ldots \ldots P(r_k) \to P(r_1 + \cdots + r_k). \tag{5.13}$$

We refer to Markl [Mar96; Mar08] for the details of the equivalence between the two definitions. Given the partial composition operations, total compositions can be defined by:

$$\gamma(p; q_1, \ldots, q_k) := (\ldots ((p \circ_k q_k) \circ_{k-1} q_{k-1}) \cdots \circ_1 q_1). \tag{5.14}$$

Conversely, if one knows the total compositions, then one can find the partial compositions by considering:

$$p \circ_i q := \gamma(p; \mathrm{id}_\mathsf{P}, \ldots, \mathrm{id}_\mathsf{P}, \underbrace{q}_{\text{position } i}, \mathrm{id}_\mathsf{P}, \ldots, \mathrm{id}_\mathsf{P}). \tag{5.15}$$

Remark 5.10 Instead of using symmetric sequences, one can use *symmetric collections*, i.e. the contravariant functors $\mathrm{Bij}^{\mathrm{op}} \to \mathsf{C}$ from the opposite of the category of finite sets and bijections to C. We refer to [Fre17a, Section 2.5] for the details. Concretely, a symmetric collection is a collection $\mathsf{P} = \{\mathsf{P}(U)\}_{U \text{ finite}}$ indexed by all finite sets and provided, for any bijection $f : U \to V$, with a morphism $f^* : \mathsf{P}(V) \to \mathsf{P}(U)$ satisfying $(g \circ f)^* = f^* \circ g^*$. The correspondence with the symmetric sequences is very simple: one can simply put $\mathsf{P}(n) = \mathsf{P}(\{1, \ldots, n\})$, and letting $\mathfrak{S}_n$ acts through the identification $\mathfrak{S}_n = \mathrm{Hom}_{\mathrm{Bij}}(\{1, \ldots, n\}, \{1, \ldots, n\})$. An operadic structure on a symmetric collection P can then be described as a unit $\mathrm{id}_\mathsf{P} \in \mathsf{P}(\{*\})$ and, for any pair $T \subset U$ of finite sets, composition operations:

$$\circ_T : \mathsf{P}(U/T) \otimes \mathsf{P}(T) \to \mathsf{P}(U), \tag{5.16}$$

where $U/T = U \setminus T \sqcup \{*\}$ is the quotient (see Definition 3.35).

Remark 5.11 There is yet another, more compact definition of operads. Let P and Q be two symmetric collections. Their composition product is defined as the symmetric collection $\mathsf{P} \circ \mathsf{Q}$ given by:

$$(\mathsf{P} \circ \mathsf{Q})(U) := \bigoplus_{r \geq 0} \mathsf{P}(\{1, \ldots, r\}) \otimes_{\mathfrak{S}_r} \left(\bigoplus_{W_1 \sqcup \cdots \sqcup W_r = U} \mathsf{Q}(W_1) \otimes \cdots \otimes \mathsf{Q}(W_r) \right), \tag{5.17}$$

where $- \otimes_{\mathfrak{S}_r} -$ denotes the coinvariants, i.e., the quotient of the tensor product by the ideal generated by the relations $x \cdot \sigma \otimes y = x \otimes \sigma \cdot y$ for $\sigma \in \mathfrak{S}_r$. The unit of this operation is the symmetric sequence I given by $\mathsf{I}(1) = \mathbb{I}$ (the unit of the monoidal product $\otimes$) and $\mathsf{I}(k) = 0$ for $k \neq 1$. Then an operad is a monoid with respect to this monoidal structure. In other words, an operad is a symmetric collection P equipped with a product $\gamma : \mathsf{P} \circ \mathsf{P} \to \mathsf{P}$ and a unit $\eta : \mathsf{I} \to \mathsf{P}$ satisfying equivariance, unitality, and associativity axioms. The associativity axiom with the current definition, combined with the unitality axiom, gives rise to the sequential and parallel composition axioms of the definition in terms of partial operations. See Markl [Mar96; Mar08] (or e.g., [LV12, Proposition 5.3.1]) for the details.

5.2.2 Algebras over an Operad

As in group theory, a central notion in operad theory is that of representations. These representations are called "algebras" in the context of operads, for a reason that will quickly become clear from the examples. Let us now define them.

Definition 5.12 Let P be an operad. An *algebra over* P (or P-algebra) is an object A equipped with a morphism of operads $\mathsf{P} \to \mathrm{End}_A$.

If we unpack the definition, we see that an algebra over P is an object A equipped with morphisms, for all $k \geq 0$:

$$\gamma_A : \mathsf{P}(k) \otimes A^{\otimes k} \to A \tag{5.18}$$

which are equivariant, unitary and associative. For $p \in \mathsf{P}(k)$ and $a_1, \ldots, a_k \in A$, let us write down:

$$p(a_1, \ldots, a_k) := \gamma_A(p \otimes a_1 \otimes \cdots \otimes a_k). \tag{5.19}$$

Then the axioms state that $(p \cdot \sigma)(a_1, \ldots, a_k) = p(a_{\sigma^{-1}(1)}, \ldots, a_{\sigma^{-1}(k)})$, that $\mathrm{id}_{\mathsf{P}}(a) = a$, and finally that:

$$(p \circ_i q)(a_1, \ldots, a_{k+l-1}) = p(a_1, \ldots, a_{i-1}, q(a_i, \ldots, a_{i+l-1}), a_{i+l}, \ldots, a_{k+l-1}). \tag{5.20}$$

These axioms are such that if we are given a tree as in Eq. (5.5) where the leaves are decorated with elements of A, then we can uniquely evaluate the tree and obtain a new element of A.

Definition 5.13 Let P be an operad and A, B two algebras over P. A *morphism of algebras* over P is a map $f : A \to B$ such that for any $p \in \mathsf{P}(k)$ and $a_1, \ldots, a_k \in A$:

$$f(p(a_1, \ldots, a_k)) = p(f(a_1), \ldots, f(a_k)). \tag{5.21}$$

Example 5.14 Let us now describe algebras over the operad As from Example 5.7. If A is such an algebra, then for any $\sigma \in \mathfrak{S}_k$ we have to define a map $\sigma : A^{\otimes k} \to A$. Given the equivariance relation, one obtains that $\sigma(a_1, \ldots, a_k) = 1_k(a_{\sigma(1)}, \ldots, a_{\sigma(k)})$ where $1_k \in \mathfrak{S}_k$ is the unit of the group. From the unitality assumption, we must have $1_1(a) = a$. Moreover, the operad structure is given by $1_k \circ_i 1_l = 1_{k+l-1}$ for any $i \in \{1, \ldots, k\}$. Applying the sequential composition axiom several times, we thus find that for any $k \geq 2$,

$$1_k(a_1, \ldots, a_k) = 1_2(a_1, 1_2(a_2, \ldots, 1_2(a_{k-1}, a_k))), \tag{5.22}$$

Moreover, if we choose any way of parenthesizing the variables $a_1, \ldots, a_k$ with the binary product 1_2, then we obtain the same element $1_k(a_1, \ldots, a_k)$. The product 1_2 thus defines an associative product on A. The elements 1_k are simply given by the iterated product. Finally, from the relation $1_2 \circ_1 1_0 = 1_2 \circ_2 1_0 = 1_1 = \mathrm{id}_{\mathsf{As}}$, we find that the image of the element 1_0 in A defines a unit for this associative product. In summary, the data of an algebra over As is equivalent to the data of a unital associative algebra (or monoid depending on the category).

Example 5.15 We can give a similar description of the operad Com. The only difference is that the associative product is invariant under permutations. Therefore, an algebra over the operad Com is a unital commutative algebra (or commutative monoid). The single element of $\mathsf{Com}(k)$ acts on algebras by multiplying k elements in any order.

Let us now get back to the general theory. As with groups, it may be useful to define an operad using a presentation by generators and relations. Informally, such a presentation is the data of: generators in each arity $k \geq 0$; an action of the symmetric group $\mathfrak{S}_k$ on the operations of arity k; and a certain number of relations between the operadic compounds of the generators (formally, an ideal in the free operad on generators). It is often more convenient to describe a presentation of an operad P by describing the category of algebras over P in terms of generating operations and relations between these operations.

Example 5.16 The operad Lie is an operad in the category of vector spaces which is generated by a binary operation λ that is antisymmetric ($\lambda \cdot (2, 1) = -\lambda$) and verifies the Jacobi relation ($\lambda \circ_1 \lambda + (\lambda \circ_1 \lambda) \cdot (2, 3, 1) + (\lambda \circ_1 \lambda) \cdot (3, 1, 2) = 0$). An algebra over this operad is precisely a Lie algebra (see Definition 2.77). The two relations can be represented graphically by the following equations:

$$
\begin{array}{ccc}
\overset{1 \quad 2}{\underset{\lambda}{\diagdown\diagup}} & = - & \overset{2 \quad 1}{\underset{\lambda}{\diagdown\diagup}} \\[-4pt]
\big| & & \big|
\end{array}
\tag{5.23}
$$

and

$$
\begin{array}{ccccccc}
\overset{2 \quad 3}{\underset{\lambda}{\diagdown\diagup}} & & & \overset{3 \quad 1}{\underset{\lambda}{\diagdown\diagup}} & & & \overset{1 \quad 2}{\underset{\lambda}{\diagdown\diagup}} \\
\overset{1 \quad \lambda}{\underset{\lambda}{\diagdown\diagup}} & + & \overset{2 \quad \lambda}{\underset{\lambda}{\diagdown\diagup}} & + & \overset{3 \quad \lambda}{\underset{\lambda}{\diagdown\diagup}} & = & 0. \\[-4pt]
\big| & & \big| & & \big|
\end{array}
\tag{5.24}
$$

Example 5.17 The operad Com has a presentation with a single binary generator $\mu \in \mathsf{Com}(2)$ which is symmetric ($\mu \cdot (2, 1) = \mu$) and associative ($\mu \circ_1 \mu = \mu \circ_2 \mu$).

Example 5.18 The operad As has a presentation with two binary generators $\mu, \bar{\mu} \in$ As(2) which are exchanged by the action of the symmetric group ($\mu \cdot (12) = \bar{\mu}$) and such that μ is associative ($\mu \circ_1 \mu = \mu \circ_2 \mu$).

Definition 5.19 Let P be an operad and X an object. The *free* P-*algebra on* X is the object given by:

$$P(X) := \bigoplus_{r \geq 0} P(r) \otimes_{\mathfrak{S}_r} X^{\otimes r}. \tag{5.25}$$

The algebra structure on P of $P(X)$ is induced by the operadic structure of P.

The terminology is justified by the following property:

Proposition 5.20 *Let* P *be an operad,* X *an object, and* A *a* P-*algebra. Then for any map* $f : X \to A$ *in the underlying category, there exists a unique morphism of* P-*algebras* $P(X) \to A$ *which coincides with* f *when restricted on generators.*

Example 5.21 Let V be a vector space. We have the following identifications:

- As(V) $= T(V)$ is the free associative algebra on V, also called the tensor algebra.
- Com(V) $= S(V)$ is the free commutative algebra on V, also called the symmetric algebra;
- Lie(V) is the free Lie algebra on V, spanned generated by Lie words on V.

Remark 5.22 Not all algebraic structures are encoded by operads. For example, there is no operad whose algebras are groups, and no operad whose algebras are fields. The problem comes from the axiom $x \cdot x^{-1} = 1$: the variable x is repeated. It is not possible to represent such a relation with the structure of an operad.

One could wonder if this is an artefact of the standard definition and if it could be possible to find an operad P such that the category of P-algebras are groups or fields. But this is not the case. For fields, note there exists free P-algebras for any operad P according to Definition 5.19. But there can be no free fields. Indeed, suppose that F were the free field on one variable x. Then there should exist a unique morphism of fields $f : F \to \mathbb{Q}$ such that $f(x) = 1$, so $x \neq 0$. There should also exist a unique morphism $g : F \to \mathbb{Q}$ such that $g(x) = 0$. This is absurd, as then $0 = g(xx^{-1}) = 1$. For groups, more elaborate arguments are needed, see e.g. [Lin13].

5.2.3   Modules over Operads

Let us now introduce the notion of modules over operads. Modules will play an important role in the study of configuration spaces. This concept generalizes the concept of algebra over an operad, in two ways: the action can be on the right (instead of always on the left for an algebra) and the elements of a module can themselves have several "inputs". Algebras are, in fact, special cases of (left) modules where all the operations have zero inputs. These modules can be used to describe functors between categories of algebras over operads (see Fresse [Fre09] and Proposition 5.37 below).

We first deal with left modules.

Definition 5.23 A *left module over an operad* P is a symmetric sequence $\mathsf{M} = \{\mathsf{M}(k)\}_{k \geq 0}$ with structural morphisms, for all $k, r_1, \ldots, r_k \geq 0$:

$$\mathsf{P}(k) \otimes \mathsf{M}(r_1) \otimes \ldots \mathsf{M}(r_k) \to \mathsf{M}(r_1 + \cdots + r_k). \tag{5.26}$$

These structural morphisms must verify axioms of equivariance, unitality and associativity similar to those of operads (stated in terms of total compositions, see Remark 5.9).

One can interpret this definition in terms of trees, just like operads: if one decorates the vertices of a tree by elements of P and the topmost vertices (i.e., those that touch leaves) by elements of M, then one can uniquely evaluate the tree and get back an element of M. Let us now give several examples of left modules.

Example 5.24 Let P be an operad and A an algebra over P. One can define a left module M_A on P by putting $\mathsf{M}_A(0) = A$ and $\mathsf{M}_A(r) = \emptyset$ is the initial object of our category C for $r \geq 0$. The structural morphisms:

$$\mathsf{P}(k) \otimes \mathsf{M}_A(0) \otimes \ldots \mathsf{M}_A(0) \to \mathsf{M}_A(0) \tag{5.27}$$

are simply the action of the operad P on its algebra A. In fact, all left modules concentrated in arity 0 are obtained as such. Moreover, the datum of a left module is equivalent to the data of a P-algebra in the category of symmetric collections, when symmetric collections are endowed with the monoidal product

$$(\mathsf{M} \otimes \mathsf{M}')(U) = \bigsqcup_{U = V \sqcup W} \mathsf{M}(V) \otimes \mathsf{M}'(W). \tag{5.28}$$

Example 5.25 Let X and Y be any two objects. One can define a left End_Y-module by $\mathrm{End}_{X,Y}(k) = \mathrm{Hom}(X^{\otimes r}, Y)$. The structural morphisms:

$$\mathrm{Hom}(Y^{\otimes k}, Y) \times \mathrm{Hom}(X^{\otimes r_1}, Y) \times \cdots \times \mathrm{Hom}(X^{\otimes r_k}, Y) \to \mathrm{Hom}(X^{\otimes (k_1 + \cdots + k_r)}, Y) \tag{5.29}$$

are given by composition of morphisms.

Example 5.26 If $f : \mathsf{P} \to \mathsf{Q}$ is an operadic morphism, then Q is a left P-module. The structural morphisms:

$$\mathsf{P}(k) \otimes \mathsf{Q}(r_1) \otimes \ldots \mathsf{Q}(r_k) \to \mathsf{Q}(r_1 + \cdots + r_k) \tag{5.30}$$

are obtained by applying the morphism f to the factor $P(k)$ and then using the operadic structure of Q.

Example 5.27 A left module over the unit operad I is simply a symmetric sequence.

The definition of a right module over an operad is similar, and we have dual kinds of examples.

Definition 5.28 A *right module over an operad* P is a symmetric sequence $M = \{M(k)\}_{k \geq 0}$ provided with of structural morphisms, for all $k, r_1, \ldots, r_k \geq 0$:

$$M(k) \otimes P(r_1) \otimes \ldots P(r_k) \to M(r_1 + \cdots + r_k). \tag{5.31}$$

This structure must satisfy equivariance, unitality, parallel and sequential composition axioms, similar to those of operads.

Example 5.29 Let X and Y be any two objects. Then $\mathrm{End}_{X,Y}$ (see Example 5.25) is a right End_X-module.

Example 5.30 If $f : P \to Q$ is an operadic morphism, then Q is a right P-module.

Example 5.31 A right module over the unit operad I is simply a symmetric sequence.

Remark 5.32 Note that right modules have another definition in terms of partial compositions. Thanks to the identity of P, the notion of right module can be defined in an equivalent way using partial compositions (compare with Remark 5.9):

$$\circ_i : M(k) \otimes P(l) \to M(k + l - 1), \text{ for } 1 \leq i \leq k, \, l \geq 0. \tag{5.32}$$

On the other hand, left modules cannot be defined using partial compositions, because that would require a unit in the left module rather than in the operad. The notion defined using partial compositions for the left modules is that of "infinitesimal" [MV09, Section 3.1], "weak" [Tur10, Definition 4.1] or "abelian" [Fre17b, Section 2.1.1] modules.

Remark 5.33 Consider the composition product defined in Remark 5.11. If we see an operad P as a monoid for this composition product, then a left module over the operad P is a left module over this monoid (in the classical sense), and a right module is a right module over this monoid.

One can also define bimodules over operads, which can be used to define functors between categories of algebras:

Definition 5.34 Let P and Q be two operads. A (P, Q)-*bimodule* is a symmetric collection M endowed with a left P-module structure and a right Q-module structure are compatible.

The prototypical example of bimodule is the following:

Example 5.35 Let X and Y be any two objects. Then $\mathrm{End}_{X,Y}$ (see Example 5.25) is an $(\mathrm{End}_Y, \mathrm{End}_X)$-bimodule.

Example 5.36 Let $f : \mathsf{P} \to \mathsf{Q}$ be a morphism of operads. Then f induces a (P, P)-bimodule structure on Q.

Of course, bimodules can be more complicated, as Remark 5.58 illustrates. One of the most interesting uses of bimodules is that they can be used to define functors between categories of algebras. We refer to Fresse [Fre09] for an extensive treatment of this point of view. We will just need the following proposition:

Proposition 5.37 *Let P and Q be two operads, and M be a (P, Q)-bimodule. There is an induced functor $\mathsf{M} \circ_{\mathsf{Q}} -$ from the category of Q-algebras to the category of P-algebras, defined by:*

$$\mathsf{M} \circ_{\mathsf{Q}} A := \bigoplus_{r \geq 0} \mathsf{M}(r) \otimes A^{\otimes r} / \sim . \tag{5.33}$$

The equivalence relation is generated by the following relation, where $m \in \mathsf{M}(r_1 + \cdots + r_k)$, $q_i \in \mathsf{Q}(r_i)$, and $a_{i,j} \in A$ (see Fig. 5.1):

$$m(q_1, \ldots, q_k) \otimes \left(\bigotimes_{i=1}^{k} \bigotimes_{j=1}^{r_i} a_{i,j} \right) \sim m \otimes \left(\bigotimes_{i=1}^{k} q_i(a_{i,1}, \ldots, a_{i,r_i}) \right). \tag{5.34}$$

The P-algebra structure on $\mathsf{M} \circ_{\mathsf{Q}} A$ is induced by the left P-action on M.

Remark 5.38 This definition extends easily to the case where A has more than one input, i.e., is itself a symmetric sequence or a module. Given a (P, Q)-bimodule M and a (Q, R)-bimodule N, we obtain a (P, R)-bimodule $\mathsf{M} \circ_{\mathsf{Q}} \mathsf{N}$ in the obvious way. This includes the case where one of the operads is the unit operad I (in which case we can simply forget about the corresponding action, which is unique).

Fig. 5.1 The relations in the composition product $\mathsf{M} \circ_{\mathsf{Q}} A$: any tree such as this one, where $m \in M(k)$, $q_i \in \mathsf{Q}(r_i)$, and $a_{i,j} \in A$, corresponds to a unique element obtained by letting the q_i's acting either on m or on the $a_{i,j}$'s

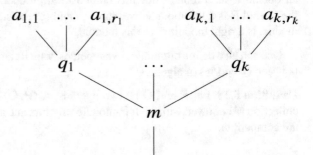

5.3   Configuration Spaces and Operads

5.3.1   Little Disks Operads

The little disks operads are a family of operads that play a central role in the theory of operads. They appeared in the first application of operads, the recognition principle, in the work of Boardman and Vogt [BV68; BV73] and May [May72] (see Theorem 5.46). We also refer to [Fre17a, Chapter 4] for a modern treatment of the subject. Since then, they have proved useful in many other applications. They also have a very strong connection with configuration spaces, as we will explain later in this section.

Definition 5.39 (See Also e.g. [Fre17a, Section 4.1])   For each dimension n, we define the *little n-disks operad*, D_n, as follows. An element $c = (c_1, \ldots, c_k) \in D_n(k)$ is a configuration of k small n-disks, with disjoint interiors, embedded in the unit n-disk $\mathbb{D}^n$. Each disk of this configuration is the image of the composite of a translation and a rescaling $c_i : \mathbb{D}^n \hookrightarrow \mathbb{D}^n$. The space of all these embedding is equipped with the compact-open topology. The action of the symmetric group reorders the disks of a configuration, and the operadic composition is given by the composition of the embeddings, as in Fig. 5.2.

Remark 5.40   An operad weakly equivalent to D_n can be defined using parallelepipeds instead of disks. We thus obtain an operad called the small n-cubes operad. More generally, any operad weakly equivalent to D_n is called an E_n-operad. There are many such operads, and we will present one in Sect. 5.3.3.

As mentioned earlier, the first application of these operads was the recognition principle of iterated loop spaces, as we now explain.

Definition 5.41   Let $X \ni *$ be a pointed topological space. Its nth *iterated loop space* of X is the space of continuous maps from the disk $\mathbb{D}^n$ that collapse the boundary to the base point, that is:

$$\Omega^n X := \left\{ \gamma : \mathbb{D}^n \to X \mid \gamma(\partial \mathbb{D}^n) = * \right\}. \tag{5.35}$$

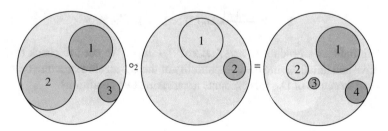

Fig. 5.2   Composition in the operad of little disks

Proposition 5.42 *The space $\Omega^n X$ is an algebra over the operad D_n.*

Proof This is almost immediate by construction. Consider a configuration $c = (c_1, \ldots, c_k) \in \mathsf{D}_n(k)$ and n-loops $\gamma_1, \ldots, \gamma_k \in \Omega^n X$. We can define $c(\gamma_1, \ldots, \gamma_k) \in \Omega^n X$ by the following formula:

$$c(\gamma_1, \ldots, \gamma_k) : \mathbb{D}^n \to X$$

$$x \mapsto \begin{cases} \gamma_i(y), & \text{if } x = c_i(y) ; \\ *, & \text{otherwise.} \end{cases} \tag{5.36}$$

It is then an easy exercise to check that the axioms of an algebra over D_n are satisfied.   □

The recognition principle says that the converse is partially true. It allows one to recognize when a space is homotopy equivalent to an iterated loop space, simply by looking at its structure. Before being able to state the theorem, we need a small preliminary.

Lemma 5.43 *Let Y be an algebra over D_n. The set $\pi_0 Y$ naturally forms a monoid, which is abelian if $n \geq 2$.*

Proof Choose any base point $c \in \mathsf{D}_n(2)$, for example this one (for $n = 2$):

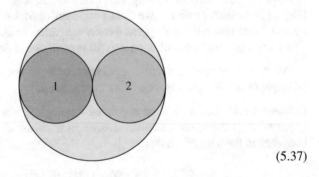

$$(5.37)$$

Using the D_n-algebra structure on Y, we get a map:

$$Y \times Y \xrightarrow{c \times 1 \times 1} \mathsf{D}_n(2) \times Y \times Y \xrightarrow{\gamma_Y} Y. \tag{5.38}$$

This map induces a product $\mu : \pi_0 Y \times \pi_0 Y \to \pi_0 Y$. We need to check that it is associative and unital. Using the compatibility of the D_n-algebra structure on Y with the operad structure of D_n, this amounts to checking two conditions.

1. For associativity, we need to check that $c \circ_1 c$ and $c \circ_2 c$ are in the same path component. This is clear:

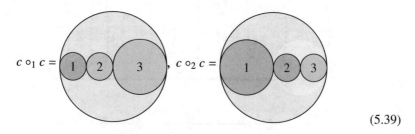

$$\tag{5.39}$$

2. For unitality, we need to check that $c \circ_1 \varnothing \in D_n(1)$ and $c \circ_2 \varnothing \in D_n(1)$ are both in the path component of the identity, where $\varnothing \in D_n(0)$ is the empty configuration. This is also obvious, as $D_n(1)$ is contractible.
3. Finally, for commutativity, we must check that c and $c \cdot (12)$ are in the same path component if $n \geq 2$, which is also obvious.

$\square$

Definition 5.44 A D_n-algebra Y is called *group-like* if the natural monoid structure on $\pi_0 Y$ is actually a group structure.

Remark 5.45 Any path-connected D_n-algebra is obviously group-like.

Theorem 5.46 (Recognition Principle, May [May72] and Boardman and Vogt [BV73]) *If Y is an algebra over D_n which is group-like, then Y has the homotopy type of an n-fold loop space.*

Since this first application, little disks operads have had many other uses. Let us mention the Deligne conjecture [KS00; MS02], which says that the Hochschild cochains $C^*(A; A)$ of an associative algebra have an action of an operad weakly equivalent to $C_*(D_2)$; the formality theorem of Hochschild cochains and its applications to quantization of Poisson manifolds [Kon99; Kon03; Tam98]; Goodwillie–Weiss manifold calculus and the calculus of embedding spaces and long knots [AT14; BW13; DH12; FTW17; FTW20; LTV10; Sin06]; factorization homology [AF15; BD04; CG17a; Lur09b; Lur17] (see Sect. 5.1).

We have already seen another example of algebra over the operad of little disks in Sect. 4.1, namely, the configuration spaces on a cylinder. In what follows, we are going to consider $X \times (0, 1)$ rather than $X \times \mathbb{R}$ in order to have easier formulas. Of course, $X \times (0, 1)$ is homeomorphic to $X \times \mathbb{R}$.

Proposition 5.47 *Let X be a topological space. The collection $\mathrm{Conf}_{X \times (0,1)}$ forms an algebra over the operad D_1 in the category of symmetric collections (or equivalent, a left D_1-module).*

Proof Let us assume that we are given:

- a finite set U and a collection of finite sets $\{V_u\}_{u \in U}$; we also denote the disjoint union by $V = \bigsqcup_{u \in U} V_u$;

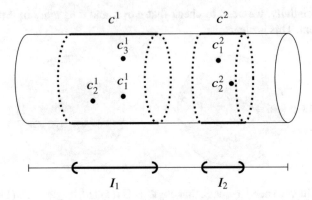

Fig. 5.3 An illustration of the D_1-algebra structure of $\mathsf{Conf}_{X \times (0,1)}$ (where $X = \mathbb{S}^1$)

- an element $I = \{I_u\}_{u \in U} \in D_1(U)$ of the little intervals operads, where $I_u :$ $[0, 1] \hookrightarrow [0, 1]$ is an affine embedding and the interiors of the I_u are pairwise disjoint;
- a collection of configurations $c = \{c^u \in \mathsf{Conf}_{X \times (0,1)}(V_u)\}_{u \in U}$, whose coordinates we denote $c^u = \{c_v^u = (x_v^u, t_v^u) \in X \times (0, 1)\}_{v \in V_u}$.

From all this data, we can define a new configuration $I(c) \in \mathsf{Conf}_{X \times \mathbb{R}}(V)$ by:

$$I(c)_v = \left(x_v^u, I_u(t_v^u)\right) \in X \times (0, 1). \tag{5.40}$$

We refer to Fig. 5.3 for a visualization of this operation.   □

The following proposition then justifies the terminology "monoid up to homotopy" that we used in Sect. 4.1:

Proposition 5.48 *The operad* D_1 *is weakly equivalent to the operad* As *of Example 5.7 (seen as a topological operad with discrete components).*

Proof Each connected component of $D_1(r)$ is easily seen to be contractible. The quotient map $D_1 \to$ As which sends a configuration of intervals to the order of the intervals' indices from left to right is a homotopy equivalence on each component.   □

5.3.2 Relationship with Configuration Spaces

The little disks operads are linked to the configuration spaces in the following way.

Lemma 5.49 *There is a homotopy equivalence* $\pi : D_n(r) \to \mathsf{Conf}_{\mathring{\mathbb{D}}^n}(r)$.

Proof The map is defined by $\pi : c \mapsto (c_1(0), \ldots, c_r(0))$, i.e., we forget the radii of the disks of a configuration and we keep only their centers. There is a map ι

in the reverse direction, which takes a configuration $(x_1, \ldots, x_r) \in \mathsf{Conf}_{\mathbb{D}^n}^\circ(r)$ to the configuration of disks $(c_1, \ldots, c_r)$ whose centers are the x_i, and such that the radius of c_i is a third of the smallest distance between x_i and the other points of the configuration. It is clear that the composition $\pi \circ \iota$ is the identity. Conversely, $\iota \circ \pi$ is homotopic to the identity: we can simply rescale the disks continuously. □

Up to homotopy, the collection of configuration spaces of $\mathbb{D}^n$ thus has an operadic structure. In what follows, we extend this to obtain a similar statement for configuration spaces of arbitrary manifolds.

Definition 5.50 Let M be a smooth compact manifold of dimension n. We denote by:

$$\mathsf{D}'_M(r) \subset \mathrm{Map}\big((\mathbb{D}^n)^{\sqcup r}, M\big) \tag{5.41}$$

the space of maps that are smooth embedding when they are restricted to each disk and such that the images of the interiors of the disks are pairwise disjoint.

Lemma 5.51 *The collection* D'_M *forms a right module over* D_n.

Proof The operadic module structure is defined using composition of embeddings. If $c = (c_1, \ldots, c_k) \in \mathsf{D}'_M(k)$ is a collection of embeddings $c_i : \mathbb{D}^n \hookrightarrow M$, and $(d_1, \ldots, d_l) \in \mathsf{D}_n(l)$ is another collection of embeddings $d_i : \mathbb{D}^n \hookrightarrow \mathbb{D}^n$, then we can define a new collection of embeddings by:

$$c \circ_i d := \big(c_1, \ldots, c_{i-1}, c_i \circ d_1, \ldots, c_i \circ d_l, c_{i+1}, \ldots, c_k\big) \in \mathsf{D}'_M(k+l-1). \tag{5.42}$$

One easily checks that this satisfies the axioms of a right module (compare with Example 5.25). □

The space $\mathsf{D}'_M(k)$ does not, however, have the homotopy type of $\mathsf{Conf}_M(k)$. The essential difference with D_n is that disks are allowed to rotate. For example, the space $\mathsf{D}_M(1)$ has the homotopy type of the frame bundle over M.

Definition 5.52 Let M be a smooth manifold of dimension n and let us note $TM = \bigcup_{x \in M} T_x M$ the tangent space of M. Then the *frame bundle* of M, Fr_M, is a $GL_n(\mathbb{R})$-principal fiber bundle on M defined by:

$$\mathrm{Fr}_M := \{(x, \xi) \mid x \in M, \ \xi \text{ basis of } T_x M\} \subset M \times (TM)^n. \tag{5.43}$$

The homotopy equivalence $\pi : \mathbb{D}'_M(1) \to \mathrm{Fr}_M$ is given on a smooth embedding $c : \mathbb{D}^n \hookrightarrow M$

$$\pi(c) = \big(c(0), (dc(0)(1, 0, \ldots, 0), \ldots, dc(0)(0, \ldots, 0, 1))\big), \tag{5.44}$$

where $dc : T\mathbb{D}^n \cong \mathbb{D}^n \times \mathbb{R}^n \to TM$ is the differential of the embedding c. More generally, we have the following statement.

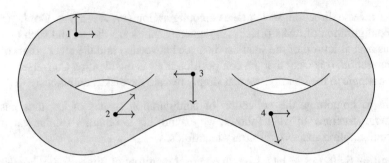

Fig. 5.4 $\mathsf{Conf}_M^{\mathrm{fr}}$: the pairs of arrows represent a basis in the tangent space at the given point

Definition 5.53 Let M be a smooth manifold. The *rth framed configuration space of M* is the space given by the pullback (see Fig. 5.4):

$$
\begin{array}{ccc}
\mathsf{Conf}_M^{\mathrm{fr}}(r) & \dashrightarrow & \mathrm{Fr}_M^r \\
\downarrow & \lrcorner & \downarrow \\
\mathsf{Conf}_M(r) & \hookrightarrow & M^r.
\end{array}
\tag{5.45}
$$

Concretely, an element of $\mathsf{Conf}_M^{\mathrm{fr}}(r)$ is the data of a configuration $x = (x_1, \ldots, x_r) \in \mathsf{Conf}_M(r)$ and a trivialization of the tangent space $T_{x_i} M$, see Fig. 5.4.

Lemma 5.54 *Let M be a smooth manifold, and $r \geq 0$ an integer. The space $\mathsf{D}'_M(r)$ has the homotopy type of the space $\mathsf{Conf}_M^{\mathrm{fr}}(r)$.*

Remark 5.55 With Campos, Ducoulombier and Willwacher [CDIW18], we have obtained a real model of the framed configuration spaces of an oriented smooth compact manifold M. This model is based on decorated graph complexes, as in the previous chapters. However, it is not as explicit as the Lambrechts–Stanley model: it depends on integrals (the partition function) that we do not know how to compute at the moment.

To obtain spaces that have the same homotopy type as the configuration spaces of M, it is therefore necessary to trivialize the tangent information.

Definition 5.56 Let M be a smooth manifold. Let us assume that M is parallelized, i.e., that its tangent fiber bundle is trivial, and let us fix an isomorphism $\tau : TM \cong M \times \mathbb{R}^n$. We can then define the subspace of the disk embeddings that respect the original parallelization:

$$
\mathsf{D}_M(r) := \big\{ c \in \mathsf{D}'_M(r) \mid \forall i, \exists \lambda > 0 \text{ s.t. } \tau(c_i(0)) \circ dc_i(0) = \lambda \, \mathrm{id}_{\mathbb{R}^n} \big\}.
\tag{5.46}
$$

Fig. 5.5 An element of
$D_2^{<3}(4)$

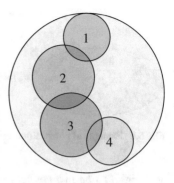

Proposition 5.57 *The space* $D_M(r)$ *has the homotopy type of* $\mathrm{Conf}_M(r)$, *and the right* D_n*-module structure on* D'_M *restricts to* D_M.

However, it is difficult to work directly with this definition. Indeed, $D_M(r)$ is not a manifold, for example, and the spaces $D_n(r)$ and $D_M(r)$ are not compact. In the next section, we'll see that we can use Fulton–MacPherson compactifications instead of D_n and D_M.

Remark 5.58 Many types of configuration spaces have links with operads. For example, consider the non-k-equal configuration spaces of Example 2.16, for some $k \geq 2$. They can be fattened to obtain a bimodule over the little disks operad, which was studied by Dobrinskaya and Turchin [DT15] and used by Ducoulombier [Duc18] to obtain delooping results. More precisely, one can introduce spaces $D_n^{<k}(r)$ given by collections of r disks $\mathbb{D}^n \subset \mathbb{D}^n$ (just like in the definition of D_n) such that the intersection of any k distinct disks is empty. Then $D_n^{<k}(r)$ has the homotopy type of $\mathrm{Conf}_{\mathbb{R}^n}^{<k}(r)$. Moreover, the collection $D_n^{<k}$ forms a (D_n, D_n)-bimodule using insertion of disks. See Fig. 5.5 for an example.

This can be generalized to any framed manifold, and one obtains a right D_n-module.

5.3.3 Operadic Structures on Compactifications

Remember that the configuration spaces of $\mathbb{R}^n$ have the same homotopy type as their Fulton–MacPherson compactifications (see Sect. 3.2). In this section, we describe an operadic structure on these compactifications, see Fig. 5.6. The operad thus obtained is moreover weakly equivalent to the operad of little n-disks, as we will see.

Proposition 5.59 . *Let* $n \geq 1$ *be an integer. The symmetric collection* $\mathsf{FM}_n = \{\mathsf{FM}_n(U)\}$ *has an operadic structure defined as follows. Let* $x \in \mathsf{FM}_n(U/T)$ *and*

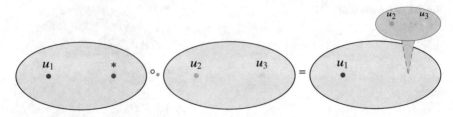

Fig. 5.6 Example of composition in FM_2

$y \in \mathsf{FM}_n(T)$ *be two configurations. Then the configuration* $x \circ_T y \in \mathsf{FM}_n(U)$ *is defined by:*

$$\theta_{ij}(x \circ_T y) := \begin{cases} \theta_{ij}(y), & \text{if } i, j \in T ; \\ \theta_{[i][j]}(x), & \text{otherwise.} \end{cases} \tag{5.47}$$

$$\delta_{ijk}(x \circ_T y) := \begin{cases} \delta_{ijk}(y), & \text{if } i, j, k \in T ; \\ 0, & \text{if } i, k \in T \text{ et } j \notin T ; \\ 1, & \text{if } i, j \in T \text{ et } k \notin T ; \\ \infty, & \text{if } j, k \in T \text{ et } i \notin T ; \\ \delta_{[i][j][k]}(x), & \text{otherwise.} \end{cases} \tag{5.48}$$

Remark 5.60 These are precisely the maps that appeared in the proof of Proposition 3.36. See Fig. 5.6 for an example.

One can easily verify that these operations induce the same operations in cohomology as those of the Theorem 5.69, but this is of course not sufficient to prove that FM_n and D_n have the same homotopy type. We do have the following result, though.

Theorem 5.61 (Salvatore [Sal19a, Proposition 3.9]) *The operads* D_n *and* FM_n *have the same homotopy type, i.e., there exists a zigzag of operadic morphisms:*

$$\mathsf{D}_n \xleftarrow{\sim} \cdot \xrightarrow{\sim} \mathsf{FM}_n \tag{5.49}$$

which are weak homotopy equivalences on each component.

Proof (Sketch of Proof) We begin by constructing an explicit "resolution" of the operad D_n using the Boardman–Vogt construction. Let us define a new operad $W\mathsf{D}_n$. The points of $W\mathsf{D}_n(k)$ are rooted decorated planar trees with k leaves. The internal vertices of these trees are decorated by elements of D_n, with an arity equal to the number of edges entering at this vertex. The inner edges of these trees are decorated by elements of the segment $[0, 1]$. We can refer to Fig. 5.7 for an example of an

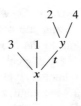

Fig. 5.7 An element of WD_n, where $x \in D_n(3)$, $y \in D_n(2)$, and $t \in [0, 1]$

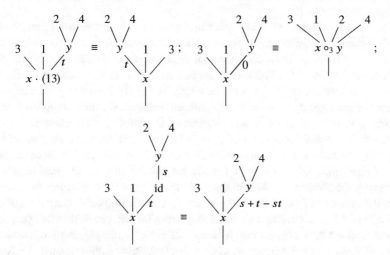

Fig. 5.8 Relations in WD_n

element. This yields a subset of the disjoint union of products of copies of $D_n(i)$ and intervals, and the topology is inherited from that larger space.

The set of these trees is then modded out by the equivalence relation generated by the following identifications, illustrated in Fig. 5.8.

- If a vertex is decorated by $x \cdot \sigma$ for $\sigma \in \mathfrak{S}_k$, then the tree is identified with the same tree where the vertex is decorated by x and the sub-trees starting from x are reordered according to σ.
- If an inner edge is decorated by $t = 0$, then the tree is identified with the same tree where the edge is contracted and the decorations of the corresponding vertices are composed using the operadic structure of D_n.
- If a vertex is decorated by id_{D_n}, then the tree is identified with the same tree where this vertex is removed. If this vertex was between two inner edges decorated respectively by s and t, then the decoration of the new edge is $s + t - st$.

This collection WD_n has an operadic structure. To compose two trees, the second one is grafted on one of the leaves of the first one and the new inner edge is decorated with $t = 1$. For example, the tree of Fig. 5.8 would be obtained as the operadic composition of two elements if $t = 1$.

There is an operadic morphism $WD_n \to D_n$ which consists simply in forgetting the decorations of the edges and using the operadic structure to compose the decorations of the vertices according to the rooted tree. One can show that this morphism is a deformation retract of topological spaces in each arity, which proves that WD_n and FM_n have the same homotopy type.

It then only remains to build a weak equivalence $\bar{\pi} : WD_n \to FM_n$. This equivalence extends the maps $\pi : D_n \to Conf_{\mathbb{R}^n} \to FM_n$ obtained by composing the map that forgets the disk radii with the quotient map onto the interior of FM_n. Let us now describe it.

Let $\mathcal{T}$ be an element of WD_n. We can represent $\mathcal{T}$ by a tree whose inner edges are decorated by positive time parameters. Let us start with the case where all t are different from 1. Then each internal edge corresponds uniquely to an affine embedding $\mathbb{D}^n \to \mathbb{D}^n$ (which corresponds to a disk in the element of $\mathbb{D}_n$ that decorates the vertex coming out of the edge). We apply a rescaling of ratio $1 - t$, where t is the edge decoration, to all these embedding. We then compose them all according to the tree to get a new element of D_n from $\mathcal{T}$. This element does not depend on the equivalence class of $\mathcal{T} \in WD_n$. We then define an element of FM_n forgetting the radii of this configuration. We can refer to the Fig. 5.9 for an example.

If $\mathcal{T}$ has inner edges decorated by 1, we cut the tree along these edges. We then apply the above procedure to each of the sub-trees thus obtained, to construct several elements of FM_n. These components are then composed using the operadic structure of FM_n, according to the starting tree. One can check that this procedure is well defined and continuous. It is by definition a morphism of operads $\bar{\pi} :$ $WD_n \to FM_n$. Since $\bar{\pi}$ extends π, since π is a homotopy equivalence, and since the

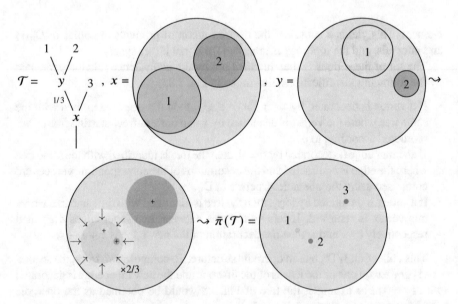

Fig. 5.9 Construction of the morphism $\bar{\pi} : WD_n \to FM_n$

inclusion $\mathsf{D}_n \to W\mathsf{D}_n$ is a homotopy equivalence, we deduce that $\bar{\pi}$ is a homotopy equivalence.   □

Remark 5.62 The same construction can be applied to FM_n to obtain an operad $W\mathsf{FM}_n$. Salvatore [Sal01; Sal19b] has shown that $W\mathsf{FM}_n$ is in fact *isomorphic* to FM_n, rather than merely homotopy equivalent. This means that the operad FM_n is cofibrant, i.e., it has a lifting property with respect to certain operadic morphisms (namely, acyclic fibrations).

Remark 5.63 In collaboration with Campos and Ducoulombier [CDI21], we have developed "leveled" versions of the Boardman–Vogt construction for operads, their modules, Hopf cooperads, and their cobimodules.

Let us now consider the configuration spaces of a fixed compact manifold without boundary M, of dimension n. As in Sect. 3.2, we choose an embedding $\iota : M \hookrightarrow \mathbb{R}^N$ for N large, which allows to define the compactification FM_M. It is further assumed that M is parallelizable, i.e., that its tangent bundle TM is trivial. Let us fix a trivialization of TM. This trivialization induces a continuous function $\tau : M \to \mathrm{Emb}(\mathbb{R}^n, \mathbb{R}^N)$ which represents the differential of ι. The map τ is such that $\tau' : (x, v) \mapsto (\iota(x), \tau(x)(v))$ makes the following diagram commute:

$$
\begin{array}{ccc}
TM & \xrightarrow{d\iota} & T\mathbb{R}^N \\
\downarrow{\scriptstyle\cong} & & \text{canon.}\downarrow{\scriptstyle\cong} \\
M \times \mathbb{R}^n & \xrightarrow{\tau'} & \mathbb{R}^N \times \mathbb{R}^N.
\end{array}
\tag{5.50}
$$

With this data, we can define a right FM_n-module structure on the collection FM_M as follows. Let $x \in \mathsf{FM}_M(U/T)$ and $y \in \mathsf{FM}_n(T)$. We consider the point $p_*(x) \in M$ (where $* \in U/T$ is the base point), which induces an embedding $\tau(p_*(x)) : \mathbb{R}^n \hookrightarrow \mathbb{R}^N$ via our parallelization τ. This embedding induces a new embedding $\mathsf{FM}_n \hookrightarrow \mathsf{FM}_N$ that is still denoted $\tau(p_*(x))$ by abuse of notation. We can therefore consider the configuration $\tau(p_*(x))(y) \in \mathsf{FM}_N(T)$. We then define a new configuration $x \circ_T y \in \mathsf{FM}_M(U)$ in the coordinates of Eq. (3.38) by:

- $p_u(x \circ_T y) := p_{[u]}(x);$

- $\theta_{ij}(x \circ_T y) := \begin{cases} \theta_{ij}\big(\tau(p_*(x))(y)\big), & \text{if } i, j \in T ; \\ \theta_{[i][j]}(x), & \text{otherwise}; \end{cases}$

- $\delta_{ijk}(x \circ_T y) := \begin{cases} \delta_{ijk}\big(\tau(p_{[T]}(x))(y)\big), & \text{if } i, j, k \in T ; \\ 0, & \text{if } i, k \in T \text{ and } j \notin T ; \\ 1, & \text{if } i, j \in T \text{ and } k \notin T ; \\ \infty, & \text{if } j, k \in T \text{ and } i \notin T ; \\ \delta_{[i][j][k]}(x), & \text{otherwise}. \end{cases}$

Roughly speaking, the virtual configuration $x \circ_T y$ is obtained by inserting the configuration $y \in \mathsf{FM}_n(T)$ at the position of $p_*(x) \in M$ in the tangent space. The parallelization is necessary to obtain coherent insertion maps at every possible position. The proof of the following proposition is similar to that of Theorem 5.61.

Proposition 5.64 *The collection* FM_M *with this structure forms a right* FM_n-*module. The pair* $(\mathsf{FM}_M, \mathsf{FM}_n)$ *is weakly equivalent to the pair* $(\mathsf{D}_M, \mathsf{D}_n)$.

Remark 5.65 The right module structure of FM_M depends on the choice of the parallelization τ. In [CDIW18, Section 5.2], with Campos, Ducoulombier, and Willwacher, we have studied the effect of a change of parallelization on graph complex models.

5.4 Models for Configuration Spaces and Their Operadic Structure

In this section, we prove that the models for configuration spaces obtained in Chap. 3 do not merely model the real homotopy types of the spaces, but that they are also compatible with the operadic structures described above.

5.4.1 Formality of the Little Disks Operads

Let us first come back to the formality of the configuration spaces of $\mathbb{R}^n$, which we discussed in Sect. 2.4. We now know that, up to homotopy, these configuration spaces have an algebraic structure: they form a (topological) operad. According to rational homotopy theory, spaces are encoded up to rational homotopy by CDGAs. In order to encode topological operads, we thus consider the following notion.

Definition 5.66 ([Fre17a, Section 3.2]) A *Hopf cooperad* is a symmetric collection $\mathsf{C} = \{\mathsf{C}(U)\}$ in the opposite category of CDGAs, equipped with a counit $\varepsilon : \mathsf{C}(\{*\}) \to \mathbb{R}$, as well as cocomposition morphisms, for any pair of finite sets $W \subset U$:

$$\circ_W^{\vee} : \mathsf{C}(U) \to \mathsf{C}(U/W) \otimes \mathsf{C}(W) \tag{5.51}$$

that satisfy equivariance, counitality, sequential cocomposition, and parallel cocomposition axioms dual to those that define operads.

Proposition 5.67 *Let* P *be a topological operad. Its homology (over any ring)* $H_*(\mathsf{P})$ *is a linear operad. Its cohomology over a field* $H^*(\mathsf{P})$ *is a Hopf cooperad if each* $\mathsf{P}(k)$ *is of finite type (i.e.* $\dim H^i(\mathsf{P}(k)) < \infty$ *for all* i, k).

Proof This stems from the functoriality of (co)homology and, over a field, from the fact that Künneth morphisms are isomorphisms. The (co)composition operations are respectively given by:

$$\circ_W : H_*(P(U/W)) \otimes H_*(P(W)) \xrightarrow{\kappa} H_*(P(U/W) \times P(W)) \xrightarrow{(\circ_W)_*} H_*(P(U)),$$
(5.52)

$$\circ_W^\vee : H^*(P(U)) \xrightarrow{\circ_W^*} H^*(P(U/W) \times P(W)) \xleftarrow[\cong]{\kappa} H^*(P(U/W)) \otimes H^*(P(W)).$$
(5.53)

It is easy to deduce that all relevant axioms are satisfied from the fact that they are satisfied for P. □

We can apply this proposition to the operad of little disks. Recall that we saw that D_1 is homotopy equivalent to the (discrete) operad As that governs topological monoids (see Proposition 5.48). For higher dimensions, we have the following theorem.

Definition 5.68 Let $n \geq 2$ be an integer. A *Poisson n-algebra* is a cochain complex A endowed by a unital commutative product and a Lie bracket of degree $1 - n$ (i.e., $\deg[x, y] = \deg x + \deg y + 1 - n$, and the antisymmetry/Jacobi relations have signs which reflect this shift) such that the bracket is a biderivation with respect to the product, i.e., $[xy, z] = x[y, z] \pm [x, z]y$. We let $Pois_n$ be the operad governing Poisson n-algebras.

Theorem 5.69 (Cohen [Coh76]) *For $n \geq 2$, the homology of D_n, the operad*

$$e_n := H_*(D_n; \mathbb{Q})$$
(5.54)

is isomorphic to $Pois_n$. The cooperad structure of its cohomology, $e_n^\vee := H^*(D_n; \mathbb{Q})$, is described as follows in the presentation of Theorem 2.88. For $T \subset U$ a pair of finite sets, cocomposition is given on generators by:

$$\circ_T^\vee : e_n^\vee(U) \to e_n^\vee(U/T) \otimes e_n^\vee(T)$$

$$\omega_{ij} \mapsto \begin{cases} 1 \otimes \omega_{ij}, & \text{if } i, j \in T ; \\ \omega_{[i][j]} \otimes 1, & \text{otherwise.} \end{cases}$$
(5.55)

Remark 5.70 Recall that the algebra $e_n^\vee$ admits a graphical interpretation (see the discussion following Theorem 2.88). An element of $e_n^\vee(U)$ can be represented by a linear combination of graphs with U vertices, modulo orientation and the Arnold relations. The cooperad structure fits into this graphical interpretation. Given a graph $[\Gamma] \in e_n^\vee(U)$ and $T \subset U$, the cocomposition $\circ_T^\vee(\Gamma) \in e_n^\vee(U/T) \otimes e_n^\vee(T)$ is the tensor $[\Gamma/T] \otimes [\Gamma_T]$, where Γ/T is the graph obtained from Γ by collapsing T to a single vertex and Γ_T is the full subgraph of Γ on the vertices T.

In order to have coherent notation, we also introduce the following notation:

$$\mathsf{e}_1 := \mathsf{As}. \tag{5.56}$$

We know from Theorem 2.101 that the configuration spaces of $\mathbb{R}^n$ are formal, i.e., their cohomology completely encodes their rational homotopy type. We moreover know that these spaces have the homotopy type of the spaces that make up the operad D_n. It is therefore natural to ask if the cohomology $\mathsf{e}_n^\vee$ encodes the rational homotopy type of the operad D_n The question must however already be clarified: indeed, Künneth morphisms are only quasi-isomorphisms at the level of cochains, not isomorphisms. So we do not have a Hopf cooperad structure on the collection $\Omega_{\mathrm{PA}}^*(\mathsf{FM}_n)$, but merely "cocomposition zigzags":

$$\Omega_{\mathrm{PA}}^*(\mathsf{FM}_n(U)) \xrightarrow{\circ_W^*} \Omega_{\mathrm{PA}}^*(\mathsf{FM}_n(U/W) \times \mathsf{FM}_n(W))$$
$$\xleftarrow{\sim} \Omega_{\mathrm{PA}}^*(\mathsf{FM}_n(U/W)) \otimes \Omega_{\mathrm{PA}}^*(\mathsf{FM}_n(W)). \tag{5.57}$$

We therefore introduce the following ad-hoc definition:

Definition 5.71 Let P be an operad in compact SA sets such that $\mathsf{P}(0)$ is a point. A *real model* of P is a Hopf cooperad C such as there are zigzags of quasi-isomorphisms of CDGAs:

$$\mathsf{C}(U) \xleftarrow{\sim} \mathsf{C}'(U) \xrightarrow{\sim} \Omega_{\mathrm{PA}}^*(\mathsf{P}(U)), \tag{5.58}$$

where C' is another Hopf cooperad, and $\mathsf{C}' \to \mathsf{C}$ is a morphism of Hopf cooperads, and the following diagrams commutes:

$$
\begin{array}{ccccc}
\mathsf{C}(U) & \xleftarrow{\ \sim\ } & \mathsf{C}'(U) & \xrightarrow{\ \sim\ } & \Omega_{\mathrm{PA}}^*(\mathsf{P}(U)) \\
\downarrow{\scriptstyle \circ_W^\vee} & & \downarrow{\scriptstyle \circ_W^\vee} & & \downarrow{\scriptstyle \circ_W^*} \\
& & & & \Omega_{\mathrm{PA}}^*(\mathsf{P}(U/W) \times \mathsf{P}(W)) \\
& & & & \sim\uparrow{\scriptstyle \kappa} \\
\mathsf{C}(U/W) \otimes \mathsf{C}(W) & \xleftarrow{\sim} & \mathsf{C}'(U/W) \otimes \mathsf{D}(W) & \xrightarrow{\sim} & \Omega_{\mathrm{PA}}^*(\mathsf{P}(U/W)) \otimes \Omega_{\mathrm{PA}}^*(\mathsf{P}(W))
\end{array}
\tag{5.59}
$$

Remark 5.72 This ad-hoc definition has theoretical foundations. Fresse [Fre17b; Fre18] developed the rational homotopy theory of operads and showed that the previous definition (with Ω_{PL}^* instead of Ω_{PA}^*) did indeed give a rational homotopy theory of topological operads which has properties similar to the rational homotopy theory of topological spaces. This theory can be adapted to the real case by replacing Ω_{PL}^* with Ω_{PA}^*, thanks to some homotopical properties that Ω_{PA}^* satisfies. There is in particular an operadic upgrade of the Ω^* (PL or PA) functor (see [Fre17b,

Section 10.1]), which produces a true Hopf cooperad $\Omega_\#^* P$ from a topological or SA operad P. Moreover, if P is a cofibrant operad, then $(\Omega_\#^* P)(r) \simeq \Omega^*(P(r))$. The restriction about $P(0)$ is necessary in order to obtain a well-behaved homotopy theory.

Definition 5.73 An operad P in compact SA sets is *formal* (over $\mathbb{R}$) if $H^*(P; \mathbb{R})$ is a real model of P.

Remark 5.74 There is a weaker notion of formality in the literature. Note that $C_*(P)$ always forms a dg-operad thanks to the Künneth morphisms. The formality that we will call "weak", as opposed to the "strong" formality defined above, requires that the dg-operads $C_*(P)$ and $H_*(P)$ be quasi-isomorphic. If an operad is strongly formal, then it is weakly formal, simply by dualizing morphisms and forgetting the Hopf structure.

Theorem 5.75 (Kontsevich [Kon99], Tamarkin [Tam03], Lambrechts and Volić [LV14], Petersen [Pet14], Fresse and Willwacher [FW20], and Boavida de Brito and Horel [BH19]) *The operad* FM_n *is formal for any n.*

This formality theorem has important consequences. We can mention in particular deformation quantization of Poisson manifolds [Kon03] and the Deligne conjecture [KS00; Tam98] (which also has several other demonstrations). It has known several variants:

- Kontsevich [Kon99] proved the weak formality on $\mathbb{R}$. We are going to review his proof below. Thanks to a later result of Guillén Santos et al. [GNPR05], weak formality on $\mathbb{R}$ turns out to be equivalent to weak formality on $\mathbb{Q}$ (a generalization of Theorem 2.96).
- Tamarkin [Tam03] showed weak formality on $\mathbb{Q}$ for $n = 2$. His proof is completely different from Kontsevich's. As mentioned in Chap. 2, the spaces $\mathsf{Conf}_{\mathbb{R}^2}(r)$ are Eilenberg–MacLane spaces and can therefore be studied completely by means of their fundamental group(oid). These are expressed in terms of braids, on which there is a "cabling" operation that describes the operadic structure. Using the existence of Drinfeld [Dri90] associators, Tamarkin deduces the formality of FM_2 on $\mathbb{Q}$. We can also refer to [Fre17b] for a refinement of this proof which allows us to deduce strong formality.
- Lambrechts and Volić [LV14] have detailed Kontsevich's proof to prove the strong formality for $n \geq 3$. We will quickly recall it below.
- Petersen [Pet14] showed weak formality for FM_2 using the action of the Grothendieck–Teichmüller group that exists on this operad. We refer to Fresse [Fre17b] for a more in-depth introduction of the Grothendieck–Teichmüller group and its relationship with the little disks operads, and Merkulov [Mer21] for a survey and a connection to graph complexes.
- Fresse and Willwacher [FW20] have shown that the operad FM_n is *intrinsically* formal on $\mathbb{Q}$ for $n \geq 3$. This means under certain technical assumptions, if P is any topological operad that has the same cohomology as FM_n, then P is formal over $\mathbb{Q}$, and is therefore rationally equivalent to FM_n.

- Boavida de Brito and Horel [BH19] have constructed an action of the Grothen-dieck–Teichmüller group on FM_n to prove the weak formality over $\mathbb{Q}$ for $n \geq 2$.

Let us now recall the proof of Kontsevich and Lambrechts–Volić. As Chap. 3 is inspired by this proof, we will settle for a sketch, the definitions and proofs being analogous.

Proof (Sketch of Proof of Theorem 5.75) As before, we are going to index our (co)operads by finite sets rather than integers. There exists a graph complex $\mathsf{Graphs}_n(U)$ that fits into a zigzag of quasi-isomorphisms:

$$\mathsf{e}_n^\vee(U) \xleftarrow{\sim} \mathsf{Graphs}_n(U) \xrightarrow{\sim} \Omega_{\mathrm{PA}}^*(\mathsf{FM}_n(U)). \tag{5.60}$$

As a vector space, $\mathsf{Graphs}_n(r)$ is generated by isomorphism classes of graphs Γ of the following type. The graph Γ has external vertices in bijection with U, and an arbitrary finite number of internal vertices. Edges are either oriented or numbered, with sign conventions similar to those in Chap. 3. The degree of Γ is $(n-1)\#E - n\#I$, where E is the set of edges and I is the set of internal vertices. The product consists in gluing graphs along their vertices, and the differential is the sum of all the ways of contracting an edge incident to an internal vertex (except the "dead ends", i.e. the edges incident to a univalent internal vertex). Finally, one mods out by graphs having components consisting entirely of internal vertices.

The morphism $\mathsf{Graphs}_n(U) \to \mathsf{e}_n^\vee(U)$ is the quotient map that sends the graphs containing internal vertices to zero. One checks that it is a quasi-isomorphism in a purely combinatorial way. The morphism $\mathsf{Graphs}_n(U) \to \Omega_{\mathrm{PA}}^*(\mathsf{FM}_n(U))$ is defined by integrals, as in Chap. 3. The "propagator" φ on $\mathsf{FM}_n(2) \cong \mathbb{S}^{n-1}$ is simply the volume form of the sphere, which is indeed a minimal form (where c_n is a constant that depends on n):

$$\varphi := c_n \sum_{i=1}^{n} (-1)^i x_i \, dx_1 \wedge \widehat{dx_i} \wedge \cdots \wedge dx_n. \tag{5.61}$$

Note in particular that for $n = 2$, $\varphi = y \, dx - x \, dy = \Im(d \log z)$ where $z = x + iy$. Compare with the Eq. (2.63).

We check that the morphism $\mathsf{Graphs}_n(U) \to \Omega_{\mathrm{PA}}^*(\mathsf{FM}_n(U))$ thus defined is a CDGA morphism, as in Chap. 3. It is clearly surjective in cohomology, so the result follows.

The new point compared to Chap. 3 is the operadic structure. We have described the cooperad structure of $\mathsf{e}_n^\vee$ in Theorem 5.69. Let us now describe the one of Graphs_n. Let $W \subset U$ be a pair of sets and $\Gamma \in \mathsf{Graphs}_n(U)$ a graph. The cocomposition $\circ_W^\vee(\Gamma)$ is a sum of several terms, indexed by all the subgraphs $\Gamma' \subset \Gamma$ whose set of external vertices is W. For such a subgraph, we define a

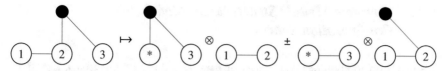

Fig. 5.10 Example of cocomposition of the form $\circ^{\vee}_{\{1,2\}}$: $\mathsf{Graphs}_n(\{1,2,3\})$ → $\mathsf{Graphs}_n(\{*,3\}) \otimes \mathsf{Graphs}_n(\{1,2\})$

quotient graph Γ/Γ' whose external vertices are U/W and whose internal vertices are those of Γ which are not in Γ'. The edges of Γ/Γ' are the edges of Γ that are not in Γ'. If one end of such an edge was a vertex of Γ', then its end in Γ/Γ' becomes the external vertex $* \in U/W$. If this procedure creates multiple edges or loops, then the result is zero. We then define an operation:

$$\circ^{\vee}_W : \mathsf{Graphs}_n(U) \to \mathsf{Graphs}_n(U/W) \otimes \mathsf{Graphs}_n(W)$$

$$\Gamma \mapsto \sum_{\substack{\Gamma' \subset \Gamma \\ V_{\text{ext}}(\Gamma')=W}} \pm \Gamma/\Gamma' \otimes \Gamma', \tag{5.62}$$

where the sign is defined by the Koszul rule. We refer to Fig. 5.10 for an example.

It is easy to check that the quotient map $\mathsf{Graphs}_n \to \mathsf{e}_n^{\vee}$ respects this operadic structure. The fact that the integration morphism $\mathsf{Graphs}_n \to \Omega^*_{\text{PA}}(\mathsf{FM}_n)$ respects it also results from a similar calculation at the level of the forms. The key point of proof lies in the decomposition of the fibered product:

$$\mathsf{FM}_n(U \sqcup I) \times_{\mathsf{FM}_n(U)} \left(\mathsf{FM}_n(U/W) \times \mathsf{FM}_n(W)\right), \tag{5.63}$$

where $\mathsf{FM}_n(U \sqcup I) \to \mathsf{FM}_n(U)$ is the canonical projection and $\mathsf{FM}_n(U/W) \times \mathsf{FM}_n(W) \to \mathsf{FM}_n(U)$ is the operadic composition. One checks that it decomposes as the union of submanifolds of $\mathsf{FM}_n(U \sqcup I)$ of codimension 1 which correspond to the various ways of deciding whether internal vertices are included in the subgraph or not. □

Remark 5.76 In Theorem 5.75, the fact that we work over a field of characteristic zero is crucial. Boavida de Brito and Horel [BH19] have obtained truncated weak formality results in characteristic $p > 2$: the operad of chains $C_*(\mathsf{D}_n; \mathbb{F}_p)$ is connected to $H_*(\mathsf{D}_n; \mathbb{F}_p)$ by a zigzag of maps that induce isomorphisms on homology up to degree $(n-1)(p-2)$. This statement is sharp, as the chain complex $C_*(\mathsf{D}_n(p); \mathbb{F}_p)$ is not $\mathfrak{S}_p$-equivariantly formal [CH18, Remark 6.9]. Salvatore [Sal19a] has shown that the little disks operads is not formal in characteristic 2 even if one forget the action of the symmetric group.

5.4.2   Operadic Module Structures on Models for Configuration Spaces

We now turn to configuration spaces of parallelized closed manifolds. Let M be such a manifold. We know that they have the homotopy type of the spaces $\mathsf{FM}_M(r)$, which form a right module on the operad FM_n as we saw in Sect. 5.3.3.

In this section, we show that the Lambrechts–Stanley model obtained in Chap. 3 is, in fact, a model for the real homotopy type of the module FM_M. We define the notion of model for a right module on an operad in a similar way to the Definition 5.71:

Definition 5.77 A *right Hopf comodule* over a Hopf cooperad C is a right comodule over C such that all the cooperad structure maps are morphisms of CDGAs. A *real model* of a pair (M, P) where P is an operad in compact SA set and M is a right module in compact SA sets is a pair (N, C) consisting of a Hopf cooperad C and a right Hopf comodule N such that there exist zigzags of quasi-isomorphisms of CDGAs:

$$\mathsf{N}(U) \leftarrow \mathsf{N}'(U) \to \Omega^*_{\mathrm{PL}}(\mathsf{N}(U)), \quad \mathsf{C}(U) \leftarrow \mathsf{C}'(U) \to \Omega^*_{\mathrm{PL}}(\mathsf{P}(U)), \qquad (5.64)$$

such that C' is a Hopf cooperad, N' is a right Hopf C'-comodule, the second zigzags makes C into a model of P and the first makes a diagram similar to the one of Definition 5.71 commute.

Proposition 5.78 *Let A be a Poincaré duality CDGA of formal dimension n. Assume that its Euler characteristic $\chi(A) = \sum_{i \geq 0}(-1)^i \dim A^i$ vanishes. Then the collection G_A of Lambrechts–Stanley models associated to A is a right Hopf comodule over the Hopf cooperad $\mathsf{e}_n^\vee$. It is equipped with the following structure maps, for $W \subset U$:*

$$\circ_W^\vee : \mathsf{G}_A(U) \mapsto \mathsf{G}_A(U/W) \otimes \mathsf{e}_n^\vee(W),$$

$$p_i^*(a) \mapsto p_{[i]}^*(a) \otimes 1 \quad \text{for } i \in U, \, a \in A,$$

$$\omega_{ij} \mapsto \begin{cases} 1 \otimes \omega_{ij}, & \text{if } i, j \in W; \\ \omega_{[i][j]} \otimes 1, & \text{otherwise.} \end{cases} \qquad (5.65)$$

Proof Let us first note the following. Let $\mathrm{vol}_A \in A^n$ be the form volume of A, i.e., the only element satisfying $\varepsilon_A(\mathrm{vol}_A) = 1$. Recall that the diagonal class Δ_A is given by $\sum_i (-1)^{|a_i|} a_i \otimes a_i^\vee$, where $\{a_i\}$ is a graded basis of A and $\{a_i^\vee\}$ is the dual basis, i.e., $\varepsilon_A(a_i a_j^\vee) = \delta_{ij} \iff a_i a_j^\vee = \delta_{ij} \mathrm{vol}_A$. Therefore, if we apply the product $\mu : A \otimes A \to A$ to the diagonal class $\Delta_A \in (A \otimes A)^n$, then we obtain the Euler class $\chi(A) \cdot \mathrm{vol}_A$. By our assumption, this element thus vanishes.

Let us now check that the structure maps written above define a right Hopf comodule. Compatibility with the product and with the cooperad structure is immediate. It only remains to check that this structure is compatible with the differential. It is clear that $d(\circ_W^\vee(p_i^*(a))) = \circ_W^\vee(d(p_i^*(a)))$ for $a \in A$ and $i \in U$. If $i \notin W$ or $j \notin W$, this relation is also clear for ω_{ij}. Finally, if $i, j \in W$, then we have:

$$d(\circ_W^\vee(\omega_{ij})) = d(1 \otimes \omega_{ij})$$
$$= 0,$$
$$\circ_W^\vee(d(\omega_{ij})) = \circ_W^\vee(p_{ij}^*(\Delta_A))$$
$$= \sum_{(\Delta_A)} p_{[i]}^*(\Delta_A') \cdot p_{[j]}^*(\Delta_A'') \otimes 1$$
$$= \chi(A) \cdot p_*^*(\mathrm{vol}_A) \otimes 1$$
$$= 0$$

□

Theorem 5.79 *Let M be a closed manifold which is simply connected, smooth, parallelizable, and of dimension $n \geq 4$. Let A be a Poincaré duality model of M. The pair $(\mathsf{G}_A, \mathsf{e}_n^\vee)$ is a real model for the pair $(\mathsf{FM}_M, \mathsf{FM}_n)$.*

Proof We simply check that the zigzag built in Chap. 3 is compatible with the comodule structure defined in Proposition 5.78. One builds a Graphs_n-Hopf comodule structure on the graph complexes which appear in this zigzag by taking as a starting point the cooperad structure of Graphs_n (see Fig. 5.10). In that process, when one contracts a subgraph, then one multiplies (together) the decorations of its vertices. It is easy to check that this structure is compatible with all morphisms of the zigzag. □

In the case of manifolds with boundary, a similar result is true (with a similar proof). We first check that we have right module structure in the same manner as in Sect. 5.3.3.

Proposition 5.80 *Let $(M, \partial M)$ be a compact manifold with boundary of dimension n. If M and $N = \partial M$ are parallelized, then the collections aFM_N and mFM_M (defined in Sect. 4.3.1) form right modules on the operad FM_n, and the action of aFM_N on mFM_M is compatible with this right module structure.*

One then only needs to check that the morphisms defined in Chap. 4 are compatible with the module structure. We then obtain the following theorems.

Theorem 5.81 ([CILW18]) *Let M be a compact manifold with boundary that satisfies the hypotheses of Theorem 4.48. Suppose furthermore that M and $N = \partial M$ are parallelized. Then with the notation of this theorem, the collections $\mathsf{aGraphs}_{A_\partial}^0$ and $\mathsf{mGraphs}_A^0$ form right Hopf comodules over Graphs_n. Moreover, the coaction of $\mathsf{aGraphs}_{A_\partial}^0$ on $\mathsf{mGraphs}_A^0$ is compatible with this comodule structure. The triple*

$(\text{mGraphs}_A^0, \text{aGraphs}_{A_\partial}^0, \text{Graphs}_n)$ *is a real model for* $(\text{mFM}_M, \text{aFM}_N, \text{FM}_n)$ *with their algebraic structures.*

Theorem 5.82 ([CILW18]) *Let M be a compact manifold with nonempty boundary satisfying the hypotheses of Theorem 4.66. With the notation of this theorem,* $\tilde{\mathsf{G}}_P$ *is a Hopf right comodule on* $\mathsf{e}_n^\vee$ *if* $\partial M \neq \varnothing$. *If moreover M is parallelized, then the pair* $(\tilde{\mathsf{G}}_P, \mathsf{e}_n^\vee)$ *is a model for the pair* $(\text{mFM}_M, \text{FM}_n)$.

5.4.3 Swiss-Cheese

Before moving on to the computation of factorization homology, let us say more about manifolds with boundary. The configuration spaces of such manifolds actually have a richer operadic structure, encoded by the Swiss-Cheese operad of Voronov [Vor99].

The Swiss-Cheese operad is an example of a colored operad. Roughly speaking, a colored operad is an operad where the inputs and the output of any operation are "colored", and operations can only be composed if colors match. Colored operads generalize operads in the same way that categories generalize monoids. In fact, a colored operad where all the operations are unary is exactly the same thing as a category.

We will not introduce colored operads in full generality, as we only need a special case of the notion. We first need the following symmetric monoidal structure on operadic right modules, which is not to be confused with the composition product of Remark 5.11.

Definition 5.83 Let P be an operad and M, M′ be right P-modules. Their tensor product, $\mathsf{M} \otimes \mathsf{M}'$, is defined for any finite set U by:

$$(\mathsf{M} \otimes \mathsf{M}')(U) := \bigoplus_{U = V \sqcup V'} \mathsf{M}(V) \otimes \mathsf{M}(V'),$$

where the direct sum runs over all partitions of U into two disjoint subsets. The symmetric collection $\mathsf{M} \otimes \mathsf{M}'$ is a right P-module using the right P-module structure of M and M′.

We then have the following compact definition that we will explain in a moment. Recall that operads can be defined in any symmetric monoidal category.

Definition 5.84 Let P be an operad. A *relative* P-*operad* is an operad in the category of right P-modules (for the symmetric monoidal structure of the previous definition).

Remark 5.85 Relative operads are also sometimes called Swiss-Cheese type operads [Wil16].

Let us now unpack this definition. Let P be an operad and Q be a relative P-operad. Then, for any finite set U, $\mathsf{Q}(U)$ is a right P-module. We denote $\mathsf{Q}(U, V) := \mathsf{Q}(U)(V)$ for a finite set V. Then we have, for any pair of sets $T \subset V$, operadic right module structure maps:

$$\circ_T : \mathsf{Q}(U, V/T) \otimes \mathsf{P}(T) \to \mathsf{Q}(U, V). \tag{5.66}$$

Moreover, the symmetric collection Q is endowed with an operad structure. This operad structure is given, for any pair $W \subset U$ and any finite sets V, T, by operadic structure maps:

$$\circ_{W,T} : \mathsf{Q}(U/W, V) \otimes \mathsf{Q}(W, T) \to \mathsf{Q}(U, V \sqcup T). \tag{5.67}$$

This definition has a graphical interpretation using trees, like in e.g., Eq. (5.4). Let us consider that operations in $\mathsf{P}(U)$ are represented by corollas with U leaves and full edges, such as this one (for $U = \{u_1, \ldots, u_k\}$ and $p \in \mathsf{P}(k)$):

$$
p = \quad
\begin{array}{c}
u_1 \quad \cdots \quad u_k \\
\diagdown \ | \ \diagup \\
p \\
|
\end{array}
\tag{5.68}
$$

Then an element of $\mathsf{Q}(U, V)$ can be represented by a corolla with two kinds of leaves: some of them are in bijection with U and are attached to full edges, while the others are in bijection with V and are attached to dashed edges. Moreover, the edge connected to the root is also dashed. We get for example a picture such as this one, for $q \in \mathsf{Q}(\{u_1, u_2\}, \{v_1, v_2, v_3\})$:

$$
q = \quad
\begin{array}{c}
u_1 \quad v_1 \quad u_2 \quad v_2 \quad v_3 \\
\diagdown \ \diagdown \ | \ \diagup \ \diagup \\
q \\
\vdots
\end{array}
\tag{5.69}
$$

Then we can graft trees as in Eq. (5.4), with the condition that full edges are grafted to full edges and dashed edges to dashed edges.

Example 5.86 Let X, Y be three objects in a symmetric monoidal category. Then the collection

$$\mathrm{End}_{X,Y}(U, V) := \mathrm{Hom}(X^{\otimes U} \otimes Y^{\otimes V}, Y) \tag{5.70}$$

forms a relative End_X-operad.

Definition 5.87 Let P be an operad and Q be a relative P-operad. A Q-*algebra* is a pair (X, Y) and a morphism of relative operads (in the obvious sense) $(Q, P) \to (\mathrm{End}_{X,Y}, \mathrm{End}_X)$.

More concretely, a Q-algebra is the data of a P-algebra X, of a unit $\mathrm{id}_Q \in Q(1, 0)$, and of maps:

$$Q(U, V) \otimes Y^{\otimes U} \otimes X^{\otimes V} \to Y \tag{5.71}$$

which satisfy obvious associativity, unitality and equivariance requirements. Note that the notion of Q-algebra depends, of course, on the operad P. What we call "the operad Q" should be more accurately called "the operad (Q, P)", but we only mention half of the structure for brevity. The context will always make clear what P is.

Let us now turn to one of the main examples of relative operads, the Swiss-Cheese operad [Vor99]. This operad is a relative D_n-operad, where D_n is the little n-disks operad of Definition 5.39. For simplicity, we see the $(n-1)$-unit disk $\mathbb{D}^{n-1}$ as embedded in the unit n-disk $\mathbb{D}^n$ as $\mathbb{D}^{n-1} \times \{0\}$. We also consider the upper half-disk to be:

$$\mathbb{D}^n_+ := \mathbb{D}^n \cap (\mathbb{D}^{n-1} \times \mathbb{R}_{\geq 0}). \tag{5.72}$$

Definition 5.88 The n-*Swiss-Cheese operad* SC_n is the relative D_n-operad defined, for finite sets U, V, as a subset $SC_n(U, V) \subset D_n(U \sqcup V)$ of configurations such that little disks indexed by U are centered on $\mathbb{D}^{n-1}$, and little disks indexed by V have their interior contained in the upper half-disk. The operadic structure is given by the composition of embeddings.

We refer to Fig. 5.11 for an illustration. It helps to refer to the disks indexed by U as "terrestrial" and the disks indexed by V as "aerial". In that picture, we represent only the upper half unit disk. Aerial disks are fully contained in that half disk, and terrestrial disks are centered on $\mathbb{D}^n \cap (\mathbb{R}^{n-1} \times \{0\})$, so the embedding is completely determined by its image in the upper part. Composition then becomes visually clear: one can either insert a configuration of full disks inside a full disk, or a configuration of full disks and half disks inside a half disk. We have drawn the second type of composition in Fig. 5.12.

The Swiss-Cheese operad also satisfies a kind of recognition principle similar to that of Theorem 5.46.

Fig. 5.11 An element of
$SC_2(\{u_1, u_2\}, \{v_1, v_2\})$

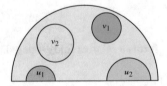

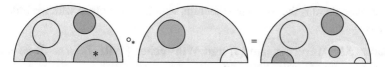

Fig. 5.12 Composition in the Swiss-Cheese operad

Definition 5.89 Let $Y \subset X$ be a pair of topological spaces and $* \in Y$ a base point.. The *relative iterated loop space* $\Omega^n(X, Y)$ is the homotopy fiber of the map $\Omega^{n-1}Y \to \Omega^{n-1}X$. More concretely, if $\mathbb{D}^n_+$ is the upper half-disk, then:

$$\Omega^n(X, Y) := \left\{ \gamma : \mathbb{D}^n_+ \to X \mid \gamma(\mathbb{D}^{n-1}) \subset Y \text{ and } \gamma(\partial \mathbb{D}^n_+ \setminus \mathbb{D}^{n-1}) = \{*\} \right\}. \tag{5.73}$$

Example 5.90 For $n = 1$, the space $\Omega^1(X, Y)$ is the homotopy fiber of the inclusion $Y \to X$. It is given by path $\gamma : [0, 1] \to X$ such that $\gamma(0) \in Y$ and $\gamma(1) = *$.

The following proposition, which is in the spirit of Proposition 5.42, is rather clear:

Proposition 5.91 *Let $* \in Y \subset X$ be any pair of pointed topological spaces. Then the pair $(\Omega^n(X, Y), \Omega^n X)$ is an SC_n-algebra.*

The recognition principle states that a kind of converse is true:

Theorem 5.92 (Hoefel et al. [HLS16], Ducoulombier [Duc14], Quesney [Que15], and Vieira [Vie20]) *Let (W, Z) be an SC_n-algebra such that W and Z are path connected. Then the pair (W, Z) is weakly homotopy equivalent to $(\Omega^n(X, Y), \Omega^n X)$ for some X, Y.*

Remark 5.93 The theorem is in fact more general than that, under some technical assumptions on W and Z that we will not detail here.

The Swiss-Cheese operad encodes, in a precise sense, the action of a D_n-algebra over a D_{n-1}-algebra. This notion was essential in, for example, deformation quantization [Kon99]. To motivate this imprecise idea into which we will note delve further, let us just mention the following result. Recall from Theorem 5.69 that an algebra over the homology of the little n-disks operad, $\mathsf{e}_n := H_*(\mathsf{D}_n)$, is an associative algebra for $n = 1$, and a Poisson n-algebra for $n \geq 2$.

Theorem 5.94 (Voronov [Vor99]) *Let $n \geq 2$ be an integer. An algebra over $H_*(\mathsf{SC}_n)$ is a triple (A, B, f) where A is an e_n-algebra, B is an e_{n-1}-algebra, and $f : A \to B$ is a central morphism of unital algebras, where central means that $[f(a), b] = 0$ for all $a \in A$ and $b \in B$ and the bracket is either the commutator (for $n \geq 2$) or part of the Poisson structure (for $n \geq 3$).*

However, there is the following theorem, which is in contrast with the little disks operads:

Theorem 5.95 (Livernet [Liv15] and Willwacher [Wil17]) *The Swiss-Cheese operad* SC_n *is not formal for any* n.

The proof of Livernet [Liv15] uses operadic Massey products, a generalization of the Massey products of Definition 2.30. The proof of Willwacher [Wil17] reduces the non-formality of SC_n to the non-formality of the obvious inclusion of operads $D_{n-1} \to D_n$, which follows from results of Turchin and Willwacher [TW18] and Fresse and Willwacher [FW20]. Note that in this theorem, the operation of arity one, i.e., the f in the description of Theorem 5.94 is used. Vieira [Vie18, Appendix A] has proved that for $n = 2$, the operad SC_2 is still not formal even when the operation of arity one is removed.

Any model of the Swiss-Cheese operad must thus have a nonzero differential. Willwacher [Wil15] has found such a model, denoted $SGraphs_n$ and based on graph complexes. Briefly, the component $SGraphs_n(U, V)$ is spanned by graphs with four types of vertices: terrestrial external vertices (in bijection with U), terrestrial internal vertices (indistinguishable, of degrees $1 - n$), aerial external vertices (in bijection with V), and aerial internal vertices (indistinguishable, of degree $-n$). In order to distinguish them, we will draw terrestrial vertices as semicircles. The edges, which are of degree $n - 1$, are directed, and their source must be aerial. Finally, if $n = 2$, then the set of terrestrial vertices is ordered. One must also consider a quotient by graphs containing only internal vertices (with a special case if $n = 2$). The following is an example of Swiss-Cheese-type graph:

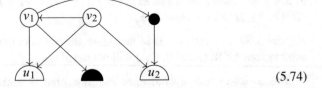

$$(5.74)$$

The most difficult part of this model is the differential. In addition to the summand which contracts edges between aerial vertices (at least one being internal), there are two extra summands, which respectively contract a subgraph containing at most one terrestrial external vertex, or contract everything outside of a subgraph containing all external vertices. The coefficients of these summands are numbers given by integrals which are in general difficult to compute. For $n = 2$, these coefficients are analogous to the ones appearing in Kontsevich's deformation quantization of Poisson manifolds [Kon03].

Now, let $(M, \partial M)$ be a manifold with boundary. Instead of plain configuration spaces, we are going to considered two-colored configuration of points in M, where some points (called terrestrial, indexed by U) belong to the boundary, and some points (called aerial, indexed by V) belong to the interior of M. Of course, as a mere topological space, this is simply the product $Conf_{\partial M}(U) \times Conf_{\mathring{M}}(V)$. However, the collection of all these configuration spaces is endowed with a richer algebraic structure. One can build compactifications $\mathsf{SFM}_M(U, V)$ of these configuration spaces in the spirit of the previous sections. If M and ∂M are framed and $TM|_{\partial M}$ splits as $T\partial M \oplus \nu_{\partial M \subset M}$, then the collection SFM_M is endowed with the structure of a right module over an operad weakly equivalent to SC_n. We arrive at the following result:

Theorem 5.96 ([CILW18]) *Let M be a framed compact manifold with framed boundary such that $TM|_{\partial M}$ splits. Then we have a real model for SFM_M based on graph complexes which is a right Hopf comodule over Willwacher's $\mathsf{SGraphs}_n$. If both M and ∂M are simply connected and $\dim M \geq 5$, then this model only depends on the real homotopy type of the inclusion $\partial M \hookrightarrow M$.*

Remark 5.97 In dimension 3, the only possible manifold is the 3-disk up to diffeomorphism thanks to the Poincaré conjecture [KL08; MT07; Per02; Per03]. Real homotopy invariance thus holds vacuously in that dimension. In dimension 4, there is only one manifold up to homeomorphism by a result of Freedman [Fre82], so the real homotopy invariance also holds vacuously for the spaces (and in fact without even the assumption on the connectivity of ∂M thanks to Corollary 4.50). However, the action of the Swiss-Cheese operad depends on the smooth structure, and there exist exotic versions of $\mathbb{R}^4$. We do not know if real homotopy invariance of the module holds in dimension 4.

Remark 5.98 There are higher-codimensional variants of the Swiss-Cheese operad. Given some fixed integers $m < n$, one can define a variant using unit disks $\mathbb{D}^n \subset \mathbb{D}^n$ of two kinds: terrestrial disks, which are centered on $\mathbb{D}^m \subset \mathbb{D}^n$, and aerial disks, which, depending on the variant, are either arbitrary [Wil17] or forbidden from touching $\mathbb{D}^m$ [Idr20]. Both variants are formal as soon as $m \leq n - 2$. This formality is related to the formality of the inclusion of operads $\mathsf{D}_m \to \mathsf{D}_n$.

5.5 Example of Calculation

Let us conclude with an example of calculation of factorization homology (see Sect. 5.1 for motivation).

We start by giving a more precise definition of factorization homology. We have defined in Remark 5.11 the composition product $\mathsf{P} \circ \mathsf{Q}$ of two symmetric collections P and Q. Let M be a parallelizable closed manifold. The collection FM_n is then a monoid for the product $\circ$ (Remark 5.11), and the collection FM_M is a right module on this monoid (Remark 5.33). An algebra A on FM_n defines a left module on FM_n concentrated in arity 0 (Example 5.24), that we will continue to denote abusively A. We thus have two structure maps:

$$\rho_M : \mathsf{FM}_M \circ \mathsf{FM}_n \to \mathsf{FM}_M, \qquad \lambda_A : \mathsf{FM}_n \circ A \to A. \tag{5.75}$$

Using these two structure maps, one can form the tensor product $\mathsf{FM}_M \circ_{\mathsf{FM}_n} A$, as defined in Proposition 5.37. This tensor product can be seen as the coequalizer of two different morphisms $\mathsf{FM}_M \circ \mathsf{FM}_n \circ A \rightrightarrows \mathsf{FM}_M \circ A$, one given by applying ρ_M, the other given by applying λ_A. Combining results from Salvatore [Sal01], Félix et al. [FHT15] and Turchin [Tur13], we obtain the following definition:

Definition 5.99 Let M be a closed framed manifold and A be an FM_n-algebra. The *factorization homology of M with coefficients in A* is the topological space given by the derived tensor product:

$$\int_M A := \mathsf{FM}_M \circ_{\mathsf{FM}_n}^{\mathbb{L}} A, \tag{5.76}$$

that is to say, the derived functor of $(M, P, N) \mapsto M \circ_P N$ (where P is an operad, M is a right P-module, and N is a left P-module).

Remark 5.100 The right module FM_M is cofibrant [Tur13, Lemma 2.3]. This property allows us to know that the derived tensor product of the Definition 5.99 is weakly equivalent to the strict tensor product, i.e., $\mathsf{FM}_M \circ_{\mathsf{FM}_n} A$. It is this strict tensor product that we describe below. With other models, however, it is important to ensure that one works with homotopically correct definitions. One possible realization of $M \circ_P^{\mathbb{L}} N$ is the bar construction, that is, the realization of the simplicial space given in degree n by:

$$B_n(M, P, N) := M \circ P^{\circ n} \circ N, \tag{5.77}$$

whose faces use either the operad structure of P or the right (resp. left) actions on M (resp. N) and whose degeneracies use the unit of P.

Remark 5.101 As we mentioned in Remark 5.1, Goodwillie–Weiss embedding calculus is adjoint to this definition of factorization homology. If M and N are closed framed manifolds such that $\dim N - \dim M \geq 3$, then $\mathrm{Emb}^{\mathrm{fr}}(M, N)$ has the homotopy type of the derived mapping space $\mathbb{R}\mathrm{Map}_{\mathsf{FM}_{\dim M}}(\mathsf{FM}_M, \mathsf{FM}_N)$. This result, just like the definition of factorization homology, can be extended to non-framed manifold by considering framed versions of configuration spaces (see Definition 5.53).

Let us now describe $\int_M A$ more concretely. This space is a quotient of the space $\mathsf{FM}_M \circ A$. The points of $\mathsf{FM}_M \circ A$ are of the type $x(a_i)_{i \in U}$, where U is a set, $x \in \mathsf{FM}_M(U)$ is a virtual configuration in M and the $a_i \in A$ are elements of A. The quotient is obtained by modding out by the relation generated by the following identifications.

- First, if $\sigma : U \to V$ is a bijection, we have an identification:

$$(x \cdot \sigma)(a_i)_{i \in U} = x(a_{\sigma^{-1}(i)})_{i \in U}. \tag{5.78}$$

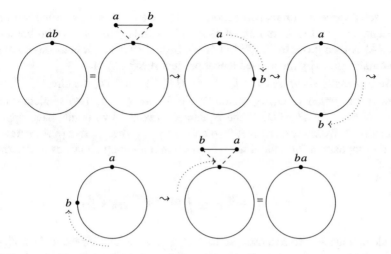

Fig. 5.13 Path in $\int_{\mathbb{S}^1} A$, where $a, b \in A$

This essentially allows us to see the points of $\mathsf{FM}_M \circ_{\mathsf{FM}_n} A$ as unordered configurations decorated with elements of A.

- Second, we have another identification, which uses the algebra structure of A and the right module structure of FM_M. For a configuration $x \in \mathsf{FM}_M(U)$, configurations $y_i \in \mathsf{FM}_n(V_i)$ (for $i \in U$) and elements $a_{i,j} \in A$ for $i \in U$ and $j \in V_i$, we have the identification:

$$(\gamma(x; y_i))(a_{i,j})_{i \in U, \, j \in V_i} = x\big(y_i (a_{i,j})_{j \in V_i}\big)_{i \in U}. \tag{5.79}$$

Graphically, this means that if points are infinitesimally close in M, then we identify the decorated configuration with the configuration obtained by replacing these infinitesimally close points by a single point decorated by the action of the infinitesimally decorated configuration (which lives in FM_n) on the corresponding elements of A. Two examples are given on Fig. 5.13.

As a set, we can thus identify $\int_M A$ with $\bigcup_{k \geq 0} \mathsf{Conf}_M(k) \times_{\mathfrak{S}_k} A^k$. The topology of the space $\int_M A$, however, is more complicated than a simple disjoint union, as we can see in the following example.

Example 5.102 Take $M = \mathbb{S}^1$ and let $x_0 \in \mathbb{S}^1$ be some base point. Let A be an associative algebra and $a, b \in A$ two elements. The algebra A becomes an algebra over FM_1 thanks to the quotient map $\mathsf{FM}_1 \xhookrightarrow{\sim} \mathsf{As} = \pi_0 \mathsf{FM}_1$. We can thus consider the space $\int_{\mathbb{S}^1} A$. In that space, there is a path between $(x_0)(ab) \in \mathsf{Conf}_M(1) \times A$ and $(x_0)(ba) \in \mathsf{Conf}_M(1) \times A$ which passes through $\mathsf{Conf}_{\mathbb{S}^1}(2) \times_{\mathfrak{S}_2} A^2$. This path is represented in Fig. 5.13.

The existence of this path is not by chance. It is known that $H_*(\int_{\mathbb{S}^1} A)$ is isomorphic to the Hochschild homology of A, which can be defined as follows.

The vector space A admits two actions of the algebra A, either on the left or on the right. This makes A into an $(A \otimes A^{\mathrm{op}})$-bimodule. The Hochschild homology $HH_*(A)$ is then given by $\mathrm{Tor}_*^{A \otimes A^{\mathrm{op}}}(A, A)$. Seen differently, it is the homology of the complex given by the derived tensor product $A \otimes_{A \otimes A^{\mathrm{op}}}^{\mathbb{L}} A$.

We can easily show that $H_*(\int_{\mathbb{S}^1} A) \simeq HH_*(A)$. Indeed, the circle $\mathbb{S}^1$ decomposes as the union $U \cup_{W \times \mathbb{R}} V$, où $U \cong V \cong D^1$ are two semicircles and $W = U \cap V \cong \mathbb{S}^0 \times D^1$ is the equator (where the two points have opposite orientations). We know that $\int_{D^1} A \simeq A$ and that $\int_{\mathbb{S}^0 \times \mathbb{R}} A \simeq \int_{\mathbb{R}} A \otimes \int_{\mathbb{R}} A \simeq A \otimes A^{\mathrm{op}}$. The decomposition formula of the factorization homology allows us to determine that:

$$\int_{\mathbb{S}^1} A \simeq \int_U A \otimes_{W \times \mathbb{R}}^{\mathbb{L}} \int_V A \simeq A \otimes_{A \otimes A^{\mathrm{op}}}^{\mathbb{L}} A.$$

Calculating the derived tensor product $A \otimes_{A \otimes A^{\mathrm{op}}}^{\mathbb{L}} A$ is, so to speak, to identify –up to homotopy– the left action and the right action of A on itself. This gives, in degree zero, the identification between the classes $[ab \otimes 1]$ and $[ba \otimes 1]$ for $a, b \in A$. But this identification is non-trivial and can give rise to non-trivial elements in higher degree, which is reflected by the existence of non-trivial paths (as above) in $\int_{\mathbb{S}^1} A$.

Remark 5.103 If A is a commutative algebra (or, more generally, a D_∞-algebra where $\mathsf{D}_\infty = \mathrm{colim}_n \mathsf{D}_n$), then it is in particular a D_n algebra using the terminal morphism $\mathsf{D}_n \to \mathsf{Com}$ (or the canonical inclusion $\mathsf{D}_n \subset \mathsf{D}_\infty$). We can thus compute the factorization homology $\int_M A$ on any parallelized manifold M. This factorization homology is then in fact given by the higher Hochschild homology of M with coefficients in A defined by Pirashvili [Pir00]. We refer to Lurie [Lur17, Theorem 5.5.3.8] or Ginot et al. [GTZ14, Theorem 5] for precise statements.

Let us now move on to a concrete computation using the results obtained in these notes. Let M be a parallelized smooth closed manifold. Let us choose a Poincaré duality model of M, which we will denote by P to distinguish it from the algebra A above. Then we know that the collection G_P formed by the Lambrechts–Stanley models of M is a model for FM_M as a right module over the operad FM_n. In particular, the pair $(\mathsf{G}_P^\vee, \mathsf{e}_n)$ (formed by the dual of G_P and the operad e_n, see Theorem 5.69) is quasi-isomorphic to the pair $(C_*(\mathsf{FM}_M), C_*(\mathsf{FM}_n))$.

Now, let A be an algebra over FM_n. Thanks to the Künneth morphism, the collection of chains $C_*(A)$ (with real coefficients) forms an algebra over the dg-operad $C_*(\mathsf{FM}_n)$. The formality of FM_n means that the operads $C_*(\mathsf{FM}_n)$ and $\mathsf{e}_n = H_*(\mathsf{FM}_n)$ are weakly equivalent. Using general theorems about operads, the homotopy category of the category of $C_*(\mathsf{FM}_n)$-algebra is equivalent to that of the category of e_n-algebras. This means that the quasi-isomorphism class of $C_*(A)$ corresponds to a single quasi-isomorphism class of e_n-algebras. Denote this class $[\tilde{A}]$, where $\tilde{A}$ is a representative e_n-algebra. The fact that the chain functor commutes with colimits implies that:

$$C_*\left(\int_M A\right) \cong C_*(\mathsf{FM}_M) \circ_{C_*(\mathsf{FM}_n)} C_*(A) \simeq \mathsf{G}_P^\vee \circ_{\mathbf{e}_n}^{\mathbb{L}} \tilde{A}. \tag{5.80}$$

This new complex, $\mathsf{G}_P^\vee \circ_{\mathbf{e}_n} \tilde{A}$, is the derived tensor product $\mathsf{G}_P^\vee \circ_{\mathbf{e}_n}^{\mathbb{L}} \tilde{A}$ defined respectively using the right action of $\mathbf{e}_n$ on $\mathsf{G}_P^\vee$, and the left action of $\mathbf{e}_n$ on $\tilde{A}$. One can explicitly describe $\mathsf{G}_P^\vee$ as a right module over $\mathbf{e}_n$ by dualizing the description of Proposition 5.78. This new complex is therefore much simpler to compute than $C_*\left(\int_M A\right)$.

Let us compute it now in a simple case to recover a theorem of Knudsen [Knu17, Theorem 3.16]. Let $\mathfrak{g}$ be a dg-Lie algebra. The free commutative graded algebra on its desuspension,

$$\tilde{A} := S\big(\mathfrak{g}[1-n]\big), \tag{5.81}$$

is an $\mathbf{e}_n$-algebra, i.e., a shifted Poisson algebra (see Theorem 5.69). The commutative product is simply the product of the free symmetric algebra. The shifted Lie bracket is extended from that of $\mathfrak{g}[1-n]$ as a biderivation, i.e., we use the following relations:

$$[a, bc] = [a, b]c + (-1)^{|b|\cdot|a|}b[a, c],$$
$$[ab, c] = a[b, c] + (-1)^{|b|\cdot|c|}[a, c]b. \tag{5.82}$$

This $\mathbf{e}_n$-algebra is a "strict" version of the universal enveloping n-algebra defined by Knudsen [Knu18b]. Indeed, we have an inclusion of operads $\iota_n : \mathsf{Lie}_n \to \mathbf{e}_n$, where Lie_n is the operad whose algebras are Lie algebras with a bracket of degree $1 - n$. Given a Lie algebra $\mathfrak{g}$, the algebra $\tilde{A} = S(\mathfrak{g}[1-n])$ is obtained by applying the extension functor $(\iota_n)_* = \mathbf{e}_n \circ_{\mathsf{Lie}_n} - \text{ to } \mathfrak{g}[1-n]$. This is the higher dimensional analog of the extension functor associated with the morphism $\iota : \mathsf{Lie} = \mathsf{Lie}_1 \to \mathsf{As} = \mathbf{e}_1$ (which sends the bracket to the commutator). This extension functor $\iota_! = \mathsf{As} \circ_{\mathsf{Lie}} -$ sends a Lie algebra $\mathfrak{g}$ on its universal enveloping algebra $\iota_!(\mathfrak{g}) = U(\mathfrak{g})$.

We can now explicitly compute $\mathsf{G}_P^\vee \circ_{\mathbf{e}_n} \tilde{A}$. Let us begin by noting that P^{-*}, the chain complex obtained by reversing the graduation of P, is a commutative graded algebra in the category of chain complexes. The tensor product $P^{-*} \otimes \mathfrak{g}$ is thus a dg-Lie algebra by setting $[a \otimes x, \ b \otimes y] = (-1)^{|b|\cdot|x|}ab \otimes [x, y]$.

Definition 5.104 Let $\mathfrak{h}$ be a dg-Lie algebra. Its Chevalley–Eilenberg complex $C_*^{CE}(\mathfrak{h})$ is given as a graded vector space by the free commutative graded algebra $S(\mathfrak{h}[-1])$ on the suspension of $\mathfrak{h}$. The differential $d_\mathfrak{h} + d_{CE}$ is the sum of the differential of $\mathfrak{h}$ (extended as a derivation) and the Chevalley–Eilenberg differential, defined by:

$$d(x_1 \otimes \cdots \otimes x_n) = \sum_{i<j} \pm x_1 \otimes \cdots \otimes x_{i-1} \otimes [x_i, x_j] \otimes x_{i+1} \otimes \cdots \otimes \hat{x}_j \otimes \cdots \otimes x_n. \tag{5.83}$$

Remark 5.105 The Chevalley–Eilenberg complex (introduced by Chevalley and Eilenberg [CE48]) can be used to define Lie algebra homology and cohomology (by considering the dual complex). If $\mathfrak{h}$ is the Lie algebra of a Lie group H, then its Lie cohomology is isomorphic to the de Rham cohomology of H. A continuous variant of the dual of the Chevalley–Eilenberg cochain complex is used to define Gelfand–Fuks cohomology (see Example 2.9).

Proposition 5.106 ([Idr19, Proposition 81]) *Let $\mathfrak{g}$ be a dg-Lie algebra, and let P be a Poincaré duality algebra of dimension $n > 1$. Let $\tilde{A} = S(\mathfrak{g}[1 - n])$ be the e_n-algebra associated to $\mathfrak{g}$. The complex $G_P^{\vee} \circ_{e_n}^{\mathbb{L}} \tilde{A}$ is quasi-isomorphic to the Chevalley–Eilenberg complex of $P^{-*} \otimes \mathfrak{g}$:*

$$\int_M [\tilde{A}] \simeq G_P^{\vee} \circ_{e_n}^{\mathbb{L}} \tilde{A} \simeq C_*^{CE}(P^{-*} \otimes \mathfrak{g}). \tag{5.84}$$

Proof (Sketch of Proof) Let us start by noting that the collection $G_P^{\vee}$ forms a right e_n-module. By restricting the structure along the inclusion $\mathsf{Lie}_n \subset e_n$, it thus forms a right Lie_n-module.

We can define a Lie algebra in the category of right Lie_n-modules by:

$$L_n := \mathsf{Lie}[1 - n] = \{\mathsf{Lie}(k)[1 - n]\}_{k \geq 0}. \tag{5.85}$$

Since P^{-*} is a graded commutative algebra, the tensor product $P^{-*} \otimes L_n = \{P^{-*} \otimes L_n(r)\}_{r \geq 0}$ remains a Lie algebra in the category of right Lie_n-modules. The proof is identical to the classical analogous result in the category of vector spaces. The Lie bracket is given by the formula $[a \otimes x, b \otimes y] = \pm ab \otimes [x, y]$ where $a, b \in P^{-*}$, $x \in L_n(r)$, $y \in L_n(s)$. We can thus define the Chevalley–Eilenberg complex $C_*^{CE}(P^{-*} \otimes L_n)$ in a way analogous to Definition 5.104.

The proof of the proposition is now very similar to that of a theorem from Félix and Thomas [FT04, Section 2]. Their result is that the spectral sequences of Bendersky–Gitler and Cohen–Taylor were dual of each other (see the end of Sect. 3.1). The Cohen–Taylor spectral sequence is the analogue of G_P, while the Bendersky–Gitler spectral sequence is the analogue of the Chevalley–Eilenberg complex. By reinterpreting Felix and Thomas' proof and using their arguments, we can show that $G_P^{\vee}$ is isomorphic, as a right Lie_n-module, to $C_*^{CE}(P^{-*} \otimes L_n)$.

The relative composition product $C_*^{CE}(P^{-*} \otimes L_n) \circ_{\mathsf{Lie}_n} \mathfrak{h}$ with a Lie_n algebra $\mathfrak{h}$ is isomorphic to $C_*^{CE}(P^{-*} \otimes \mathfrak{h})[1 - n]$. Using the decomposition $e_n = \mathsf{Com} \circ \mathsf{Lie}_n$ in terms of the operads Com and Lie_n by a distributive law (cf. [LV12, Section 13.3]), we obtain that:

$$G_P^{\vee} \circ_{e_n} S(\mathfrak{h}[1 - n]) \cong G_P^{\vee} \circ_{\mathsf{Lie}_n} \mathfrak{h}[1 - n] \cong C_*^{CE}(P^{-*} \otimes \mathfrak{h}). \tag{5.86}$$

To compute the *derived* composition product of the proposition, we need to replace $\mathfrak{h}$ with a cofibrant resolution of $\mathfrak{g}$ in the above expression. However, the Chevalley-Eilenberg complex preserves quasi-isomorphisms, so this does not change the type of quasi-isomorphism. We thus obtain the desired result.   □

Proposition 5.106 thus allows us to compute the chains of factorization homology over a simply connected, parallelized, smooth closed manifold of dimension ≥ 4, with coefficients in a universal enveloping n-algebra.

Remark 5.107 As the functor Ω^*_{PA} sends colimits to limits, we can also compute the real homotopy type of $\int_M A$. The result is simply easier to state in the dual framework.

In the previous proof, we described $G^\vee_P$ as a right Lie_n-module. However, this collection actually forms a module on $e_n = \mathsf{Com} \circ \mathsf{Lie}_n$. The right Com-module structure has been described in [Idr18, Section 4.4] and we will recall it here. We only need to describe the action of the generators: the unit $\eta \in \mathsf{Com}(0)$ and the product $\mu \in \mathsf{Com}(2)$.

To describe these actions, let us begin by noting the following facts. The isomorphism of Poincaré duality, $P^\vee \cong P^{n-*}$, induces a cocommutative product $\Delta : P^{n-*} \rightarrow (P^{n-*})^{\otimes 2}$ which is dual to the product of P. Moreover, the augmentation $\varepsilon : P^{n-*} \rightarrow \mathbb{R}$ is dual to the unit of P with this point of view.

Example 5.108 Let $P = H^*(\mathbb{S}^2 \times \mathbb{S}^2) = S(a,b)/(a^2, b^2)$ (where $\deg a = \deg b = 2$) and set $ab = v$ with $\varepsilon(v) = 1$. Then $\Delta(1) = v \otimes 1 + 1 \otimes v + a \otimes b + b \otimes a$, $\Delta(a) = a \otimes v + v \otimes a$ (similarly for b) and $\Delta(v) = v \otimes v$.

Let us now describe the action of Com on $G^\vee_P$. Recall that Com is generated by the operations $\eta \in \mathsf{Com}(0)$ (the unit) and $\mu \in \mathsf{Com}(2)$ (the product), so we only need to describe the action of these elements to fully describe the right module structure. Let U be a set and $U = \bigsqcup_{i \in I} V_i$ a partition of U. We will denote by a dot ("·") the tensor product in $P \otimes \mathsf{Lie}_n(V_i)$ to distinguish it from the other tensor products. Let

$$X = \bigotimes_{i \in U} a_i \cdot \lambda_i \in G^\vee_P(U) \tag{5.87}$$

be an element, where $a_i \in P$ and $\lambda_i \in \mathsf{Lie}_n(V_i)$ is a Lie word. Let also $j \in V_i \subset U$ be an index.

1. Let us first describe $X \circ_j \eta \in G^\vee_P(U \setminus \{j\})$, i.e., the insertion of the unit in one of the inputs of X. If λ_i has at least one bracket (i.e., $\#V_i \geq 2$), then $X \circ_j \eta$ vanishes. Otherwise, we have $V_i = \{j\}$ and $\lambda_i = \bar{\lambda}_i$ id is a scalar multiple of the identity (for $\bar{\lambda}_i \in \mathbb{R}$). In this case, we have:

$$X \circ_j \eta = \bar{\lambda}_i \varepsilon(a_i) \bigotimes_{i' \neq i} a_{i'} \cdot \lambda_{i'}. \tag{5.88}$$

2. Let us now describe $X \circ_j \mu \in G^\vee_P(U \sqcup \{j', j''\} \setminus \{j\})$, i.e., the insertion of the product $\mu \in \mathsf{Com}(2) = \mathsf{Com}(\{j', j''\})$ in the jth input, where j' and j'' are

two new input names. Using the distributive law $\mathsf{Lie}_n \circ \mathsf{Com} \to \mathsf{Com} \circ \mathsf{Lie}_n$ the defines the n-Poisson operad, the element $\lambda_i \circ_j \mu$ is a sum:

$$\lambda_i \circ_j \mu = \sum_{V_i = V_i' \sqcup V_i''} \mu(\underbrace{\lambda_i'}_{\in \mathsf{Lie}_n(V_i' \sqcup \{j'\})} , \underbrace{\lambda_i''}_{\in \mathsf{Lie}_n(V_i'' \sqcup \{j''\})}). \tag{5.89}$$

Let us also write $\Delta(a_i) = \sum_{(a_i)} a_i' \otimes a_i''$ for the coproduct of a. Then we have:

$$X \circ_j \mu = \sum_{V_i = V_i' \sqcup V_i''} \sum_{(a)} \bigotimes_{k \neq i \in U} (a_k \cdot \lambda_k) \otimes a_i' \cdot \lambda_i' \otimes a_i'' \cdot \lambda_i''. \tag{5.90}$$

Example 5.109 Consider the element $X = a \cdot \mathrm{id}(u_1) \otimes b \cdot \lambda(u_2, u_3) \in G_P^\vee(\{u_1, u_2, u_3\})$, where $a, b \in P^{-*}$, $\mathrm{id}(u_1) \in \mathsf{Lie}_n(\{u_1\})$ is the identity, and $\lambda(u_2, u_3) \in \mathsf{Lie}_n(\{u_2, u_3\})$ is the Lie bracket.

1. We have $X \circ_{u_1} \eta = \varepsilon(a) b \cdot \lambda(u_2, u_3)$ and $X \circ_{u_2} \eta = X \circ_{u_3} \eta = 0$.
2. Let us use Sweedler's notation for co-products: $\Delta(a) = \sum_{(a)} a' \otimes a'' \in (P^{n-*})^{\otimes 2}$ and the same for $\Delta(b)$. Let us write $\mu \in \mathsf{Com}(\{v_1, v_2\})$ for the product. Then we see that:

$$X \circ_{u_1} \mu = \sum_{(a)} a' \cdot \mathrm{id}(v_1) \otimes a'' \cdot \mathrm{id}(v_2) \otimes b \cdot \lambda(u_2, u_3),$$

$$X \circ_{u_2} \mu = \sum_{(b)} a \cdot \mathrm{id}(u_1) \otimes \big(b' \cdot \mathrm{id}(v_1) \otimes b'' \cdot \lambda(v_2, u_3) \pm b' \cdot \lambda(v_1, u_3) \otimes b'' \cdot \mathrm{id}(v_2) \big),$$

and $X \circ_{u_3} \mu$ has a description similar to $X \circ_{u_2} \mu$.

Remark 5.110 In [Idr18], using the e_n-module structure above, we have extended the calculation of the Proposition 5.78 to the case where $\mathfrak{g}$ is a "unitary" Lie algebra, i.e., when it is equipped with an element c satisfying $[c, x] = 0$ for any $x \in \mathfrak{g}$. The analog of the universal enveloping n-algebra is then the quotient $S(\mathfrak{g}[1 - n])/(c = 1)$. We find a result analogous to the one of Proposition 5.106.

References

[ADK19]   B.H. An, G.C. Drummond-Cole, B. Knudsen, Subdivisional spaces and graph braid groups. Documenta Math. **24**, 1513–1583 (2019). ISSN: 1431-0635. arXiv: 1708.02351

[AK04]   M. Aouina, J.R. Klein, On the homotopy invariance of configuration spaces. Algebr. Geom. Topol. **4**, 813–827 (2004). ISSN: 1472-2747. https://doi.org/10.2140/agt.2004. 4.813. arXiv: math/0310483

[Arn69]   V.I. Arnold, The cohomology ring of the colored braid group. Math. Notes **5**(2), 138–140 (1969). ISSN: 0025-567X. https://doi.org/10.1007/BF01098313

[AT14]   G. Arone, V. Turchin, On the rational homology of high dimensional analogues of spaces of long knots. Geom. Topol. **18**(3), 1261–1322 (2014). ISSN: 1465-3060. https://doi.org/10.2140/gt.2014.18.1261. arXiv: 1105.1576

[AT15]   G. Arone, V. Turchin, Graph-complexes computing the rational homotopy of high dimensional analogues of spaces of long knots. Ann. Inst. Fourier **65**(1), 1–62 (2015). https://doi.org/10.5802/aif.2924. arXiv: 1108.1001

[Art47]   E. Artin, Theory of braids. Ann. Math. 2nd ser. **48**(1), 101–126 (1947). ISSN: 0003-486X. https://doi.org/10.2307/1969218

[AS92]   S. Axelrod, I.M. Singer, Chern–Simons perturbation theory, in *Proceedings of the XXth International Conference on Differential Geometric Methods in Theoretical Physics* (New York, June 3–7, 1991), vol. 1, ed. by S. Catto, A. Rocha (World Scientific Publishing, River Edge, 1992), pp. 3–45. https://doi.org/10.1142/1537. arXiv: hep-th/9110056.pdf

[AS94]   S. Axelrod, I.M. Singer, Chern–Simons perturbation theory II. J. Differential Geom. **39**(1), 173–213 (1994). ISSN: 0022-040X. https://doi.org/10.4310/jdg/1214454681. arXiv: hep-th/9304087

[AF15]   D. Ayala, J. Francis, Factorization homology of topological manifolds. J. Topol. **8**(4), 1045–1084 (2015). ISSN: 1753-8416. https://doi.org/10.1112/jtopol/jtv028. arXiv: 1206.5522

[AFT17]   D. Ayala, J. Francis, H.L. Tanaka, Factorization homology of stratified spaces. Selecta Math. New ser. **23**(1), 293–362 (2017). ISSN: 1022-1824. https://doi.org/10.1007/ s00029-016-0242-1. arXiv: 1409.0848

[BD04]   A. Beilinson, V. Drinfeld, *Chiral Algebras*. American Mathematical Society Colloquium Publications, vol. 51 (American Mathematical Society, Providence, 2004), 375 pp. ISBN: 0-8218-3528-9. https://doi.org/10.1090/coll/051

© The Author(s), under exclusive license to Springer Nature Switzerland AG 2022   173
N. Idrissi, *Real Homotopy of Configuration Spaces*, Lecture Notes
in Mathematics 2303, https://doi.org/10.1007/978-3-031-04428-1

[Bel04]   P. Bellingeri, On presentations of surface braid groups. J. Algebra **274**(2), 543–
            563 (2004). ISSN: 0021-8693. https://doi.org/10.1016/j.jalgebra.2003.12.009. arXiv:
            math/0110129

[BG91]   M. Bendersky, S. Gitler, The cohomology of certain function spaces. Trans. Amer.
            Math. Soc. **326**(1), 423–440 (1991). ISSN: 0002-9947. https://doi.org/10.2307/
            2001871

[BMP05]   B. Berceanu, M. Markl, Ş. Papadima. Multiplicative models for configuration spaces
            of algebraic varieties. Topology **44**(2), 415–440 (2005). ISSN: 0040-9383. https://doi.
            org/10.1016/j.top.2004.10.002. arXiv: math/0308243

[Bir69]   J.S. Birman, On braid groups. Comm. Pure Appl. Math. **22**, 41–72 (1969) ISSN: 0010-
            3640. https://doi.org/10.1002/cpa.3160220104

[BV68]   J.M. Boardman, R.M. Vogt, Homotopy-everything H-spaces. Bull. Amer. Math. Soc.
            74, 1117–1122 (1968). ISSN: 0002-9904. https://doi.org/10.1090/S0002-9904-1968-
            12070-1

[BV73]   J.M. Boardman, R.M. Vogt, *Homotopy Invariant Algebraic Structures on Topological
            Spaces*. Lecture Notes in Mathematics, vol. 347 (Springer, Berlin, 1973), 257 pp.
            ISBN: 978-3-540-06479-4. https://doi.org/10.1007/BFb0068547

[BCT89]   C.-F. Bödigheimer, F. Cohen, L. Taylor, On the homology of configuration spaces.
            Topology **28**(1), 111–123 (1989). ISSN: 0040-9383. https://doi.org/10.1016/0040-
            9383(89)90035-9

[BM20]   M. Bökstedt, E. Minuz, Cohomology of generalised configuration spaces of points on
            $\mathbb{R}^r$ 2020. arXiv: 2004.08370. Pre-published

[BC98]   R. Bott, A.S. Cattaneo, Integral invariants of 3-manifolds. J. Differ. Geom. **48**(1), 91–
            133 (1998). ISSN: 0022-040X. https://doi.org/10.4310/jdg/1214460608

[BT82]   R. Bott, L.W. Tu, *Differential Forms in Algebraic Topology*. Graduate Texts in
            Mathematics, vol. 82 (Springer, New York, 1982), 331 pp. ISBN: 978-1-4419-2815-3.
            https://doi.org/10.1007/978-1-4757-3951-0

[Cor15]   H.C. Bulens, Rational model of the configuration space of two points in a simply
            connected closed manifold. Proc. Amer. Math. Soc. **143**(12), 5437–5453 (2015).
            ISSN: 0002-9939. https://doi.org/10.1090/proc/12666. arXiv: 1505.06290

[CLS18]   H.C. Bulens, P. Lambrechts, D. Stanley, Ra-tional models of the complement of a
            subpolyhedron in a manifold with boundary. Can. J. Math **70**(2), 265–293 (2018).
            https://doi.org/10.4153/CJM-2017-021-3. arXiv: 1505.04816

[CLS19]   H.C. Bulens, P. Lambrechts, D. Stanley, Pretty rational models for Poincaré duality
            pairs. Algebr. Geom. Topol. **19**, 1–30 (2019). https://doi.org/10.2140/agt.2019.19.1.
            arXiv: 1505.04818

[CW16]   R. Campos, T. Willwacher, A model for configuration spaces of points (2016). arXiv:
            1604.02043. Pre-published

[CDIW18]   R. Campos, J. Ducoulombier, N. Idrissi, T. Willwacher, A model for framed configu-
            ration spaces of points (2018). arXiv: 1807.08319. Pre-published

[CILW18]   R. Campos, N. Idrissi, P. Lambrechts, T. Willwacher, Configuration spaces of
            manifolds with boundary (2018). arXiv: 1802.00716. Pre-published

[CIW19]   R. Campos, N. Idrissi, T. Willwacher, Configuration spaces of surfaces (2019). arXiv:
            1911.12281. Pre-published

[CDI21]   R. Campos, J. Ducoulombier, N. Idrissi, Boardman–Vogt resolutions and
            bar/cobar constructions of (co)operadic (co)bimodules. High. Struct. (2021). arXiv:
            1911.09474. Forthcoming

[CM10]   A.S. Cattaneo, P. Mnëv, Remarks on Chern-Simons invariants. Comm. Math.
            Phys. **293**(3), 803–836 (2010). ISSN: 0010-3616. https://doi.org/10.1007/s00220-
            009-0959-1. arXiv: 0811.2045

[CE48]   C. Chevalley, S, Eilenberg, Cohomology theory of Lie groups and Lie algebras.
            Trans. Amer. Math. Soc. **63**, 85–124 (1948). ISSN: 0002-9947. https://doi.org/10.
            2307/1990637

[Chu12]   T. Church, Homological stability for configuration spaces of manifolds. Invent. Math. **188**(2), 465–504 (2012). ISSN: 0020-9910. https://doi.org/10.1007/s00222-011-0353-4. arXiv: 1103.2441

[CF13]   T. Church, B. Farb, Representation theory and homological stability. Adv. Math. **245**, 250–314 (2013). ISSN: 0001-8708. https://doi.org/10.1016/j.aim.2013.06.016. arXiv: 1008.1368

[CEF15]   T. Church, J.S. Ellenberg, B. Farb, FI-modules and stability for representations of symmetric groups. Duke Math. J. **164**(9), 1833–1910 (2015). ISSN: 0012-7094. https://doi.org/10.1215/00127094-3120274. arXiv: 1204.4533

[CH18]   J. Cirici, G. Horel, étale cohomology, purity and formality with torsion coefficients (2018). arXiv: 1806.03006. Pre-published

[Coh73a] F. Cohen, Cohomology of braid spaces. Bull. Amer. Math. Soc. **79**, 763–766 (1973). ISSN: 0002-9904. https://doi.org/10.1090/S0002-9904-1973-13306-3

[Coh73b] M.M. Cohen, *A Course in Simple-Homotopy Theory*. Graduate Texts in Mathematics, vol. 10 (Springer, New York, 1973), 144 pp.

[Coh76]   F.R. Cohen, The homology of $\mathscr{C}_{n+1}$ spaces, $n \geq 0$, in *The Homology of Iterated Loop Spaces*, ed. by F.R. Cohen, T.J. Lada, J.P. May. Lecture Notes in Mathematics, vol. 533 (Springer, Berlin, 1976). Chap. 3, pp. 207–351. ISBN: 978-3-540-07984-2. https://doi.org/10.1007/BFb0080467

[CT78]   F.R. Cohen, L.R. Taylor, Computations of Gelfand–Fuks cohomology, the cohomology of function spaces, and the cohomology of configuration spaces, in *Geometric Applications of Homotopy Theory I* (Evanston, IL, Mar 21–26, 1977), ed. by M.G. Barratt, M.E. Mahowald. Lecture Notes in Mathematics, vol. 657 (Springer, Berlin, 1978), pp. 106–143. https://doi.org/10.1007/BFb0069229

[CG17a]   K. Costello, O. Gwilliam, *Factorization Algebras in Quantum Field Theory*, vol. 1. New Mathematical Monographs, vol. 31 (Cambridge University Press, Cambridge, 2017), 387 pp. ISBN: 978-1-107-16310-2. https://doi.org/10.1017/9781316678626

[CG17b]   K. Costello, O. Gwilliam, *Factorization Algebras in Quantum Field Theory*, vol. 2 (Cambridge University Press, Cambridge, 2017). https://people.math.umass.edu/~gwilliam/vol2may8.pdf

[BH19]   P.B. de Brito, G. Horel, On the formality of the little disks operad in positive characteristic (2019). arXiv: 1903.09191. Pre-published

[BW13]   P.B. de Brito, M. Weiss, Manifold calculus and homotopy sheaves. Homology Homotopy Appl. **15**(2), 361–383 (2013). ISSN: 1532-0073. https://doi.org/10.4310/HHA.2013.v15.n2.a20. arXiv: 1202.1305

[BW18]   P.B. de Brito, M.S. Weiss, Spaces of smooth embeddings and configuration categories. J. Topol. **11**(1), 65–143 (2018). ISSN: 1753-8416. https://doi.org/10.1112/topo.12048. arXiv: 1502.01640

[Del75]   P. Deligne, Poids dans la cohomologie des variétés algébriques, in *Proceedings of the International Congress of Mathematicians* (Vancouver, 1974), vol. 1 (1975), pp. 79–85. ISBN: 0-919558-04-6. https://www.mathunion.org/icm/proceedings

[DGMS75] P. Deligne, P. Griffiths, J. Morgan, D. Sullivan, Real homotopy theory of Kähler manifolds. Invent. Math. **29**(3), 245–274 (1975). ISSN: 0020-9910. https://doi.org/10.1007/BF01389853

[DV10]   A. Djament, C. Vespa, Sur l'homologie des groupes orthogonaux et symplectiques à coefficients tordus. Ann. Sci. Éc. Norm. Supér. 4th ser. **43**(3), 395–459 (2010). ISSN: 0012-9593. https://doi.org/10.24033/asens.2125

[DV17]   A. Djament, C. Vespa, Foncteurs faiblement polynomiaux. Int. Math. Res. Not. (2017). https://doi.org/10.1093/imrn/rnx099. arXiv: 1308.4106

[DT15]   N. Dobrinskaya, V. Turchin, Homology of non-k-overlapping discs. Homology Homotopy Appl. **17**(2), 261–290 (2015). ISSN: 1532-0073. https://doi.org/10.4310/HHA.2015.v17.n2.a13. arXiv: 1403.0881

[DW15]   V. Dolgushev, T. Willwacher, Operadic twisting. With an application to Deligne's conjecture. J. Pure Appl. Algebra **219**(5), 1349–1428 (2015). ISSN: 0022-4049. https://doi.org/10.1016/j.jpaa.2014.06.010

[Dri90]   V.G. Drinfeld, On quasitriangular quasi-Hopf algebras and on a group that is closely connected with Gal($\overline{\mathbb{Q}}/\mathbb{Q}$). Algebra i Analiz **2**(4), 149–181 (1990). ISSN: 0234-0852

[Duc14]   J. Ducoulombier, Swiss-cheese action on the totalization of operads under the monoid actions actions operad (2014). arXiv: 1410.3236. Pre-published

[Duc18]   J. Ducoulombier, Delooping of high-dimensional spaces of string links (2018). arXiv: 1809.00682

[DH12]   W. Dwyer, K. Hess, Long knots and maps between operads. Geom. Topol. **16**(2), 919–955 (2012). ISSN: 1465-3060. https://doi.org/10.2140/gt.2012.16.919. arXiv: 1006.0874

[DS95]   W.G. Dwyer, J. Spaliński, Homotopy theories and model categories, in *Handbook of Algebraic Topology*, ed. by I.M. James (North-Holland, Amsterdam, 1995), pp. 73–126. https://doi.org/10.1016/B978-044481779-2/50003-1

[FN62a]   E. Fadell, L. Neuwirth, Configuration spaces. Math. Scand. **10**, 111–118 (1962). ISSN: 0025-5521. https://doi.org/10.7146/math.scand.a-10517

[Far03]   M. Farber, Topological complexity of motion planning. Discrete Comput. Geom. **29**(2), 211–221 (2003). ISSN: 0179-5376. https://doi.org/10.1007/s00454-002-0760-9. arXiv: math/0111197

[Far18]   M. Farber, Configuration spaces and robot motion planning algorithms, in *Combinatorial and Toric Homotopy*. Lecture Notes Series, Institute for Mathematical Sciences, National University of Singapore, vol. 35 (World Scientific Publication, Hackensack, 2018), pp. 263–303. arXiv: 1701.02083

[FT04]   Y. Félix, J.-C. Thomas, Configuration spaces and Massey products. Int. Math. Res. Not. **33**, 1685–1702 (2004). ISSN: 1073-7928. https://doi.org/10.1155/S1073792804140270. arXiv: math/0304226

[FHT01]   Y. Félix, S. Halperin, J.-C. Thomas, *Rational Homotopy Theory*. Graduate Texts in Mathematics, vol. 205 (Springer, New York, 2001), 535 pp. ISBN: 0-387-95068-0. https://doi.org/10.1007/978-1-4613-0105-9

[FOT08]   Y. Félix, J. Oprea, D. Tanré, *Algebraic Models in Geometry*. Oxford Graduate Texts in Mathematics, vol. 17 (Oxford University Press, Oxford, 2008), 460 pp. ISBN: 978-0-19-920651-3

[FHT15]   Y. Félix, S. Halperin, J.-C. Thomas, *Rational Homotopy Theory II* (World Scientific, Hackensack, 2015), 412 pp. ISBN: 978-981-4651-42-4. https://doi.org/10.1142/9473

[FN62b]   R. Fox, L. Neuwirth, The Braid Groups. Math. Scand. **10**, 119–126 (1962). ISSN: 0025-5521. https://doi.org/10.7146/math.scand.a-10518

[Fra13]   J. Francis, The tangent complex and Hochschild cohomology of $\mathscr{E}_n$-rings. Compos. Math. **149**(3), 430–480 (2013). ISSN: 0010-437X. https://doi.org/10.1112/S0010437X12000140

[Fre82]   M.H. Freedman, The topology of four-dimensional manifolds. J. Differ. Geom. **17**(3), 357–453 (1982). ISSN: 0022-040X. https://doi.org/10.4310/jdg/1214437136

[Fre09]   B. Fresse, *Modules over Operads and Functors*. Lecture Notes in Mathematics, vol. 1967 (Springer, Berlin, 2009), 308 pp. ISBN: 978-3-540-89055-3. https://doi.org/10.1007/978-3-540-89056-0. arXiv: 0704.3090

[Fre17a]   B. Fresse, *Homotopy of Operads and Grothendieck–Teichmüller Groups*, vol. 1. The Algebraic Theory and its Topological Background. Mathematical Surveys and Monographs, 217 (Americal Mathematical Society, Providence, 2017), 532 pp. ISBN: 978-1-4704-3481-6

[Fre17b]   B. Fresse, *Homotopy of Operads and Grothendieck–Teichmüller Groups*, vol. 2. The Applications of (Rational) Homotopy Theory Methods. Mathematical Surveys and Monographs, 217 (Americal Mathematical Society, Providence, 2017), 704 pp. ISBN: 978-1-4704-3482-3

[Fre18]   B. Fresse, The extended rational homotopy theory of operads. Georgian Math. J. **25**(4), 493–512 (2018). ISSN: 1072-947X. https://doi.org/10.1515/gmj-2018-0061. arXiv: 1805.00530

[FW20]   B. Fresse, T. Willwacher, The intrinsic formality of E_n-operads. J. Eur. Math. Soc. **22**(7), 2047–2133 (2020). https://doi.org/10.4171/JEMS/961. arXiv: 1503.08699

[FTW17]   B. Fresse, V. Turchin, T. Willwacher, The rational homotopy of mapping spaces of E_n operads (2017). arXiv: 1703.06123. Pre-published

[FTW20]   B. Fresse, V. Turchin, T. Willwacher, On the rational homotopy type of embedding spaces of manifolds in $\mathbb{R}^n$ (2020). arXiv: 2008.08146. Pre-published

[FM94]   W. Fulton, R. MacPherson, A compactification of configuration spaces. Ann. Math. 2nd ser. **139**(1), 183–225 (1994). ISSN: 0003-486X. https://doi.org/10.2307/2946631

[GZ67]   P. Gabriel, M. Zisman, *Calculus of Fractions and Homotopy Theory*. Ergebnisse der Mathematik und ihrer Grenzgebiete, vol. 35 (Springer, New York, 1967), 168 pp. ISBN: 978-3-642-85846-8. https://doi.org/10.1007/978-3-642-85844-4

[Get95]   E. Getzler, Operads and moduli spaces of genus 0 Riemann surfaces, in *The Moduli Space of Curves* (Texel Island, 1994). Progress in Mathematics, vol. 129 (Birkhäuser, Boston, 1995), pp. 199–230. arXiv: alg-geom/9411004

[GJ94]   E. Getzler, J.D.S. Jones, Operads, homotopy algebra and iterated integrals for double loop spaces (1994). arXiv: hep-th/9403055. Pre-published

[Ghr01]   R. Ghrist, Configuration spaces and braid groups on graphs in robotics, in *Knots, Braids, and Mapping Class Groups*. Papers dedicated to Joan S. Birman (New York, 1998), ed. by J. Gilman, W.M. Menasco, X.-S. Lin. AMS/IP Studies in Advanced Mathematics, vol. 24 (American Mathematical Society, Providence, 2001), pp. 29–40. ISBN: 978-0-8218-2966-0. https://doi.org/10.1090/amsip/024. arXiv: math/9905023

[Gin15]   G. Ginot, Notes on factorization algebras, factorization homology and applications, in *Mathematical Aspects of Quantum Field Theories*, ed. by D. Calaque, T. Strobl. Mathematical Physics Studies (Springer, Cham, 2015), pp. 429–552. ISBN: 978-3-319-09948-4. https://doi.org/10.1007/978-3-319-09949-1_13. arXiv: 1307.5213

[GTZ14]   G. Ginot, T. Tradler, M. Zeinalian, Higher Hochschild homology, topological chiral homology and factorization algebras. Comm. Math. Phys. **326**(3), 635–686 (2014). ISSN: 0010-3616. https://doi.org/10.1007/s00220-014-1889-0

[GK94]   V. Ginzburg, M. Kapranov, Koszul duality for operads. Duke Math. J. **76**(1), 203–272 (1994). ISSN: 0012-7094. https://doi.org/10.1215/S0012-7094-94-07608-4

[GJ99]   P.G. Goerss, J.F. Jardine, *Simplicial Homotopy Theory*. Progress in Mathematics, vol. 174 (Birkhäuser, Basel, 1999), 510 pp. ISBN: 3-7643-6064-X. https://doi.org/10.1007/978-3-0348-8707-6

[GW99]   T.G. Goodwillie, M. Weiss, Embeddings from the point of view of immersion theory. Part II. Geom. Topol. **3**, 103–118 (1999). https://doi.org/10.2140/gt.1999.3.103. arXiv: math/9905203

[GM13]   P. Griffiths, J. Morgan, *Rational Homotopy Theory and Differential Forms*, 2nd ed. Progress in Mathematics, vol. 16 (Springer, New York, 2013), 224 pp. ISBN: 978-1-4614-8467-7. https://doi.org/10.1007/978-1-4614-8468-4

[Gro19]   K. Grossnickle, Non-k-equal configuration and immersion spaces. PhD thesis. Kansas State University (2019)

[GJ15]   J. Guaschi, D. Juan-Pineda, A survey of surface braid groups and the lower algebraic K-theory of their group rings, in *Handbook of Group Actions*, vol. 2. Advanced Lectures in Mathematics 32 (International Press, Somerville, 2015), pp. 23–75. arXiv: 1302.6536

[Hae76]   A. Haefliger, Sur la cohomologie de l'algèbre de Lie des champs de vecteurs. Ann. Sci. École Norm. Sup. 4th ser. **9**(4), 503–532 (1976). ISSN: 0012-9593. https://doi.org/10.24033/asens.1316

[Háj20]   P. Hájek, Hodge decompositions and Poincare duality models (2020). arXiv: 2004.07362. Pre-published

[HLTV11] R. Hardt, P. Lambrechts, V. Turchin, I. Volić, Real homotopy theory of semi-algebraic sets. Algebr. Geom. Topol. **11**(5), 2477–2545 (2011). ISSN: 1472-2747. https://doi. org/10.214/agt.2011.11.2477. arXiv: 0806.0476

[Hat02]   A. Hatcher, *Algebraic Topology* (Cambridge University Press, Cambridge, 2002), 544 pp. ISBN: 0-521-79160-X. http://pi.math.cornell.edu/~hatcher/AT/AT.pdf

[Hes07]   K. Hess, Rational homotopy theory. A brief introduction, in *Interactions Between Homotopy Theory and Algebra* (University of Chicago, July 26–Aug. 6, 2004), ed. by L.L. Avramov, J.D. Christensen, W.G. Dwyer, M.A. Mandell, B.E. Shipley. Contemporary Mathematics, vol. 436 (Americal Mathematical Society, Providence, 2007), pp. 175–202. https://doi.org/10.1090/conm/436/08409

[Hoe12]   E. Hoefel, Some elementary operadic homotopy equivalences, in *Topics in Noncommutative Geometry*. Clay Mathematics Proceedings, vol. 16 (Americal Mathematical Society, Providence, 2012), pp. 67–74. arXiv: 1110.3116

[HLS16]   E. Hoefel, M. Livernet, J. Stasheff, A_∞-actions and recognition of relative loop spaces. Topology Appl. **206**, 126–147 (2016). ISSN: 0166-8641. https://doi.org/10. 1016/j.topol.2016.03.023. arXiv: 1312.7155

[Hov99]   M. Hovey, *Model Categories*. Mathematical Surveys and Monographs, vol. 63 (American Mathematical Society, Providence, 1999), 209 pp. ISBN: 0-8218-1359-5

[Hur91]   A. Hurwitz, Ueber Riemann'sche Flächen mit gegebenen Verzwei-gungspunkten. Math. Ann. **39**(1), 1–60 (1891). ISSN: 0025-5831. https://doi.org/10.1007/ BF01199469

[Idr18]   N. Idrissi, Curved koszul duality for algebras over unital operads (2018). arXiv: 1805.01853. Pre-published

[Idr19]   N. Idrissi, The Lambrechts–Stanley model of configuration spaces. Invent. Math **216**(1), 1–68 (2019). ISSN: 1432-1297. https://doi.org/10.1007/s00222-018-0842-9. arXiv: 1608.08054

[Idr20]   N. Idrissi, Formality of a higher-codimensional Swiss-Cheese operad. Algebr. Geom. Topol. (2020). arXiv: 1809.07667. Forthcoming

[KS16]   S. Kallel, I. Saihi, Homotopy groups of diagonal complements. Algebr. Geom. Topol. **16**(5), 2949–2980 (2016). ISSN: 1472-2747. https://doi.org/10.2140/agt.2016. 16.2949

[KL08]   B. Kleiner, J. Lott, Notes on Perelman's papers. Geom. Topol. **12**(5), 2587–2855 (2008). ISSN: 1465-3060. https://doi.org/10.2140/gt.2008.12.2587. arXiv: math/0605667

[Knu17]   B. Knudsen, Betti numbers and stability for configuration spaces via factorization homology. Algebr. Geom. Topol. **17**(5), 3137–3187 (2017). ISSN: 1472-2747. https:// doi.org/10.2140/agt.2017.17.3137. arXiv: 1405.6696

[Knu18a] B. Knudsen, Configuration spaces in algebraic topology (2018). arXiv: 1803.11165

[Knu18b] B. Knudsen, Higher enveloping algebras. Geom. Topol. **22**(7), 4013–4066 (2018). ISSN: 1465-3060. https://doi.org/10.2140/gt.2018.22.4013. arXiv: 1605.01391

[Koh85]   T. Kohno, Série de Poincaré–Koszul associée aux groupes de tresses pures. Invent. Math. **82**(1), 57–75 (1985). ISSN: 0020-9910. https://doi.org/10.1007/BF01394779

[Kon93]   M. Kontsevich, Formal (non)commutative symplectic geometry, in *The Gelfand Mathematical Seminars. 1990–1992*, ed. by I.M. Gelfand, L. Corwin, J. Lepowsky (Birkhäuser, Boston, 1993), pp. 173–187. ISBN: 978-0-8176-3689-0. https://doi.org/ 10.1007/978-1-4612-0345-2_11

[Kon99]   M. Kontsevich, Operads and motives in deformation quantization. Lett. Math. Phys. **48**(1), 35–72 (1999). ISSN: 0377-9017. https://doi.org/10.1023/A:1007555725247. arXiv: math/9904055

[Kon03]   M. Kontsevich, Deformation quantization of Poisson manifolds. Lett. Math. Phys. **66**(3), 157–216 (2003). ISSN: 0377-9017. https://doi.org/10.1023/B:MATH. 0000027508.00421.bf. arXiv: q-alg/9709040

[KS00]   M. Kontsevich, Y. Soibelman, Deformations of algebras over operads and the Deligne conjecture, in *Quantization, Deformation, and Symmetries*. Conférence Moshé Flato

(Dijon, Sept. 5–8, 1999), ed. by G. Dito, D. Sternheimer, vol. 1. Mathematical Physics Studies 21 (Kluwer Academic Publishers, Dordrecht, 2000), pp. 255–307. arXiv: math/0001151

[Kri94]   I. Kriz, On the rational homotopy type of configuration spaces. Ann. Math. 2nd ser. **139**(2), 227–237 (1994). ISSN: 0003-486X. https://doi.org/10.2307/2946581

[LS04]   P. Lambrechts, D. Stanley, The rational homotopy type of configuration spaces of two points. Ann. Inst. Fourier **54**(4), 1029–1052 (2004). ISSN: 0373-0956. https://doi.org/ 10.5802/aif.2042

[LS08a]   P. Lambrechts, D. Stanley, A remarkable DGmodule model for configuration spaces. Algebr Geom. Topol. **8**(2), 1191–1222 (2008). ISSN: 1472-2747. https://doi.org/10. 214/agt.2008.8.1191. arXiv: 0707.2350

[LS08b]   P. Lambrechts, D. Stanley, Poincaré duality and commutative differential graded algebras. Ann. Sci. éc. Norm. Sup. 4th ser. **41**(4), 495–509 (2008). ISSN: 0012-9593. https://doi.org/10.24033/asens.2074. arXiv: math/0701309

[LTV10]   P. Lambrechts, V. Turchin, I. Volić, The rational homology of spaces of long knots in codimension >2. Geom. Topol. **14**(4), 2151–2187 (2010). ISSN: 1465-3060. https:// doi.org/10.2140/gt.2010.14.2151. arXiv: math/0703649

[LV14]   P. Lambrechts, I. Volić, Formality of the little N-disks operad. Mem. Amer. Math. Soc. **230**(1079) (2014). ISSN: 0065-9266. https://doi.org/10.1090/memo/1079. arXiv: 0808.0457

[Lam00]   S. Lambropoulou, Braid structures in knot complements, handlebodies and 3-manifolds, in *Knots in Hellas* (Delphi, 1998), vol. 24. Ser. Knots Everything (World Scientific Publishing, River Edge, 2000), pp. 274–289. https://doi.org/10. 1142/9789812792679_0017

[Laz55]   M. Lazard, Lois de groupes et analyseurs. Ann. Sci. éc. Norm. Sup. 3rd ser. **72**, 299–400 (1955). ISSN: 0012-9593. https://doi.org/10.24033/asens.1038

[Lev95]   N. Levitt, Spaces of arcs and configuration spaces of manifolds. Topology **34**(1), 217–230 (1995). ISSN: 0040-9383. https://doi.org/10.1016/0040-9383(94)E0012-9

[Lin13]   Z. Lin, *Are Groups Algebras over an Operad?* Answer on Math Stack Exchange (2013). https://math.stackexchange.com/a/366371 (visited on 02/04/2021)

[Liv15]   M. Livernet, Non-formality of the Swiss-cheese operad. J. Topol. **8**(4), 1156–1166 (2015). https://doi.org/10.1112/jtopol/jtv018. arXiv: 1404.2484

[LV12]   J.-L. Loday, B. Vallette, *Algebraic Operads.* Grundlehren der mathematischen Wissenschaften, vol. 346 (Springer, Berlin, 2012), 634 pp. ISBN: 978-3-642-30361-6. https://doi.org/10.1007/978-3-642-30362-3

[LS05]   R. Longoni, P. Salvatore, Configuration spaces are not homotopy invariant. Topology **44**(2), 375–380 (2005). ISSN: 0040-9383. https://doi.org/10.1016/j.top.2004.11.002. arXiv: math/0401075

[Lur09a]   J. Lurie, Derived algebraic geometry VI. $\mathbb{E}[k]$-algebras (2009). arXiv: 0911.0018

[Lur09b]   J. Lurie, On the classification of topological field theories, in *Current Developments in Mathematics 2008*, ed. by D. Jerison, B. Mazur, T. Mrowka, W. Schmid, R.P. Stanley, S.-T. Yau (International Press, Somerville, 2009), pp. 129–280. https://doi. org/10.4310/CDM.2008.v2008.n1.a3. arXiv: 0905.0465

[Lur17]   J. Lurie, Higher algebra (2017). http://www.math.harvard.edu/~lurie/papers/HA.pdf

[LS34]   L. Lusternik, L. Schnirelmann, *Methodes topologiques dans les problemes variationnels.* Exposés sur l'analyse mathématique et ses applications, vol. 3 (Hermann, Paris, 1934)

[Mac65]   S. Mac Lane, Categorical algebra. Bull. Amer. Math. Soc. **71**, 40–106 (1965). ISSN: 0002-9904. https://doi.org/10.1090/S0002-9904-1965-11234-4

[Mag74]   W. Magnus, Braid groups. A survey, in *Proceedings of the Second International Conference on the Theory of Groups* (Australian National University, Aug. 13–24, 1973), ed. by M.F. Newman. Lecture Notes in Mathematics, vol. 372 (Springer, Berlin, 1974), pp. 463–487. https://doi.org/10.1007/978-3-662-21571-5_49

[Mar96]   M. Markl, Models for operads. Comm. Algebra **24**(4), 1471–1500 (1996). ISSN:
            0092-7872. https://doi.org/10.1080/00927879608825647
[Mar08]   M. Markl, Operads and PROPs, in *Handbook of Algebra*, ed. by Michiel Hazewinkel,
            vol. 5 (Elsevier/North-Holland, Amsterdam, 2008), pp. 87–140. https://doi.org/10.
            1016/S1570-7954(07)05002-4
[Mas58]   W.S. Massey, Some higher order cohomology operations, in *Symposium internacional
            de topología algebraica* (Mexico City, Aug. 1958). Universidad Nacional Autónoma
            de México and UNESCO (1958), pp. 145–154
[May72]   J.P. May, *The Geometry of Iterated Loop Spaces*. Lecture Notes in Mathematics,
            vol. 271(Springer, Berlin, 1972), 175 pp. ISBN: 978-3-540-05904-2. https://doi.org/
            10.1007/BFb0067491
[McC01]   J. McCleary, *A User's Guide to Spectral Sequences*, 2nd edn. Cambridge Studies
            in Advanced Mathematics, vol. 58 (Cambridge University Press, Cambridge, 2001),
            561 pp. ISBN: 0-521-56759-9
[MS02]   J.E. McClure, J.H. Smith, A solution of Deligne's Hochschild cohomology conjecture,
            in *Recent Progress in Homotopy Theory* (Baltimore, MD, Mar. 17–27, 2000), ed.
            by D.M. Davis, J. Morava, G. Nishida, W.S. Wilson, N. Yagita. Contemporary
            Mathematics, vol. 293 (Americal Mathematical Society, Providence, 2002), pp. 153–
            193. https://doi.org/10.1090/conm/293/04948
[McD75]   D. McDuff, Configuration spaces of positive and negative particles. Topology **14**, 91–
            107 (1975). ISSN: 0040-9383. https://doi.org/10.1016/0040-9383(75)90038-5
[Mer21]   S. Merkulov, Grothendieck–Teichmüller group, operads and graph complexes. A
            survey, in *Integrability, Quantization, and Geometry*, vol. 2. Quantum Theories and
            Algebraic Geometry, ed. by S. Novikov, I. Krichever, O. Ogievetsky, S. Shlosman.
            AMS Proceedings of Symposia in Pure Mathematics 103 (2021), pp. 383–446. arXiv:
            1904.13097
[MV09]   S. Merkulov, B. Vallette, Deformation theory of representations of prop(erad)s. I. J.
            Reine Angew. Math. **634**, 51–106 (2009). ISSN: 0075-4102. https://doi.org/10.1515/
            CRELLE.2009.069
[MS74]   J.W. Milnor, J.D. Stasheff, *Characteristic Classes*. Annals of Mathematics Studies,
            vol. 76. (Princeton University Press, Princeton; Universtiy Tokyo Press, Tokyo, 1974),
            331 pp. ISBN: 978-0-6910-8122-9
[Mor78]   J.W. Morgan, The algebraic topology of smooth algebraic varieties. Publ. Math. Inst.
            Hautes études Sci. **48**, 137–204 (1978). ISSN: 0073-8301. http://www.numdam.org/
            item?id=PMIHES_1978__48__137_0
[MT07]   J. Morgan, G. Tian, *Ricci flow and the Poincaré conjecture*. Clay Mathematics
            Monographs, vol. 3 (American Mathematical Society, Providence; Clay Mathematics
            Institute, Cambridge, 2007), 521 pp. ISBN: 978-0-8218-4328-4. arXiv: math/0607607
[Mor01]   S. Morita, *Geometry of Characteristic Classes*. Translations of Mathematical Mono-
            graphs, vol. 199 (American Mathematical Society, Providence, 2001), 185 pp. ISBN:
            0-8218-2139-3. Trans. of Japanese. Iwanami Series in Modern Mathematics, 1999
[MW12]   S. Morrison, K. Walker, Blob homology. Geom. Topol. **16**(3), 1481–1607 (2012).
            ISSN: 1465-3060. https://doi.org/10.2140/gt.2012.16.1481
[Nas52]   J. Nash, Real algebraic manifolds. Ann. Math. 2nd ser. **56**, 405–421 (1952). ISSN:
            0003-486X. https://doi.org/10.2307/1969649
[OT92]   P. Orlik, H. Terao, *Arrangements of Hyperplanes*. Grundlehren der Mathematischen
            Wissenschaften, vol. 300 (Springer, Berlin, 1992), 325 pp. ISBN: 3-540-55259-6
[Per02]   G. Perelman, The entropy formula for the Ricci flow and its geometric applications
            (2002). arXiv: math/0211159
[Per03]   G. Perelman, Ricci flow with surgery on three-manifolds (2003). arXiv: math/0303109
[Pet14]   D. Petersen, Minimal models, GT-action and formality of the little disk operad. Selecta
            Math. New ser. **20**(3), 817–822 (2014). ISSN: 1022-1824. https://doi.org/10.1007/
            s00029-013-0135-5. arXiv: 1303.1448

[Pet20]   D. Petersen, Cohomology of generalized configuration spaces. Compos. Math. **156**(2), 251–298 (2020). ISSN: 0010-437X. https://doi.org/10.1112/s0010437x19007747. arXiv: 1807.07293

[Pir00]   T. Pirashvili, Hodge decomposition for higher order Hochschild homology. Ann. Sci. éc. Norm. Sup. 4th ser. **33**(2), 151–179 (2000). ISSN: 0012-9593. https://doi.org/10.1016/S0012-9593(00)00107-5

[PP05]   A. Polishchuk, L. Positselski, *Quadratic Algebras*. University Lecture Series, vol. 37 (American Mathematical Society, Providence, 2005), 159 pp. ISBN: 0-8218-3834-2

[Que15]   A. Quesney, Swiss Cheese type operads and models for relative loop spaces (2015). arXiv: 1511.05826. Pre-published

[Qui67]   D.G. Quillen, *Homotopical Algebra*. Lecture Notes in Mathematics, vol. 43 (Springer, Berlin, 1967), 156 pp. https://doi.org/10.1007/BFb0097438

[Qui69]   D. Quillen, Rational homotopy theory. Ann. Math. 2nd ser. **90**, 205–295 (1969). ISSN: 0003-486X. https://doi.org/10.2307/1970725

[RS18]   G. Raptis, P. Salvatore, A remark on configuration spaces of two points. Proc. Edinb. Math. Soc. 2nd ser. **61**(2), 599–605 (2018). ISSN: 0013-0915. https://doi.org/10.1017/s0013091517000384. arXiv: 1604.01558

[Sal01]   P. Salvatore, Configuration spaces with summable label, in *Cohomological Methods in Homotopy Theory*. Barcelona Conference on Algebraic Topology (Bellaterra, Spain, June 5–10, 1998), ed. by J. Aguadé, C. Broto, C. Casacuberta. Progress in Mathematics, vol. 196 (Birkhäuser, Basel, 2001), pp. 375–395. ISBN: 978-3-0348-9513-2. https://doi.org/10.1007/978-3-0348-8312-2_23. arXiv: math/9907073

[Sal19a] P. Salvatore, Planar non-formality of the little discs operad in characteristic two. Q. J. Math. **2**, 689–701 (2019). ISSN: 0033-5606. https://doi.org/10.1093/qmath/hay063. arXiv: 1807.11671

[Sal19b] P. Salvatore, The Fulton MacPherson operad and the W-construction (2019). arXiv: 1906.07696. Pre-published

[GNPR05] F.G. Santos, V. Navarro, P. Pascual, A. Roig, Moduli spaces and formal operads. Duke Math. J. **129**(2), 291–335 (2005). ISSN: 0012-7094. https://doi.org/10.1215/S0012-7094-05-12924-6

[Seg73]   G. Segal, Configuration-spaces and iterated loop-spaces. Invent. Math. **21**, 213–221 (1973). ISSN: 0020-9910. https://doi.org/10.1007/BF01390197

[Seg79]   G. Segal, The topology of spaces of rational functions. Acta Math. **143**(1–2), 39–72 (1979). ISSN: 0001-5962. https://doi.org/10.1007/BF02392088

[Seg04]   G. Segal, The definition of conformal field theory, in *Topology, Geometry and Quantum Field Theory*. London Mathematical Society Lecture Note Series, vol. 308 (Cambridge University Press, Cambridge, 2004), pp. 421–577.

[Ser53]   J.-P. Serre, Groupes d'homotopie et classes de groupes abéliens. Ann. Math. 2nd ser. **58**, 258–294 (1953). ISSN: 0003-486X. https://doi.org/10.2307/1969789

[Sin04]   D.P. Sinha, Manifold-theoretic compactifications of configuration spaces. Selecta Math. New ser. **10**(3), 391–428 (2004). ISSN: 1022-1824. https://doi.org/10.1007/s00029-004-0381-7. arXiv: math/0306385

[Sin06]   D.P. Sinha, Operads and knot spaces. J. Amer. Math. Soc. **19**(2), 461–486 (2006). ISSN: 0894-0347. https://doi.org/10.1090/S0894-0347-05-00510-2

[Sin13]   D.P. Sinha, The (non-equivariant) homology of the little disks operad, in *Proceedings of the School and Conference*. Operads 2009 (Luminy, Apr. 20–30, 2009), ed. by J.-L. Loday, B. Vallette. Séminar Congress, vol. 26 (Soc. Math. France, Paris, 2013), pp. 253–279. ISBN: 978-2-85629-363-8. arXiv: math/0610236

[Sna74]   V.P. Snaith, A stable decomposition of $\Omega^n S^n X$. J. London Math. Soc. 2nd ser. **7**, 577–583 (1974). ISSN: 0024-6107. https://doi.org/10.1112/jlms/s2-7.4.577

[Sta61]   J.D. Stasheff, Homotopy associativity of H-spaces. PhD Thesis. Princeton University (1961), 116 pp.

[Sul77]   D. Sullivan, Infinitesimal computations in topology. Publ. Math. Inst. Hautes études
            Sci. **47**, 269–331 (1977). ISSN: 0073-8301. http://www.numdam.org/item/PMIHES_
            1977__47__269_0

[Tam51]   D. Tamari, Monoïdes préordonnés et chaînes de Malcev. Ph.D. Thesis. Université de
            Paris (1951)

[Tam98]   D.E. Tamarkin, Another proof of M. Kontsevich formality theorem (1998). arXiv:
            math/9803025

[Tam03]   D.E. Tamarkin, Formality of chain operad of little discs. Lett. Math. Phys. **66**(1–
            2), 65–72 (2003). ISSN: 0377-9017. https://doi.org/10.1023/B:MATH.0000017651.
            12703.a1

[Tog73]   A. Tognoli, Su una congettura di Nash. Ann. Scuola Norm. Sup. Pisa. 3rd ser. **27**,
            167–185 (1973). http://www.numdam.org/item/ASNSP_1973_3_27_1_167_0

[Tot96]   B. Totaro, Configuration spaces of algebraic varieties. Topology **35**(4), 1057–1067
            (1996). ISSN: 0040-9383. https://doi.org/10.1016/0040-9383(95)00058-5

[Tur10]   V. Turchin, Hodge-type decomposition in the homology of long knots. J. Topol. **3**(3),
            487–534 (2010). ISSN: 1753-8416. https://doi.org/10.1112/jtopol/jtq015

[Tur13]   V. Turchin, Context-free manifold calculus and the Fulton–MacPherson operad.
            Algebr. Geom. Topol. **13**(3), 1243–1271 (2013). ISSN: 1472-2747. https://doi.org/
            10.2140/agt.2013.13.1243. arXiv: 1204.0501

[TW18]   V. Turchin, T. Willwacher, Relative (non-)formality of the little cubes operads and
            the algebraic Cerf Lemma. Am. J. Math. **140**(2), 277–316 (2018). ISSN: 0002-9327;
            1080-6377/e. https://doi.org/10.1353/ajm.2018.0006. arXiv: 1409.0163

[Vie18]   R.V. Vieira, Princípio de reconhecimento de espaços de laços relativos. Portuguese.
            Ph.D. Thesis. Universidade de São Paulo (2018)

[Vie20]   R.V. Vieira, Relative recognition principle. Algebr. Geom. Topol. **20**(3), 1431–1486
            (2020). https://doi.org/10.2140/agt.2020.20.1431. arXiv: 1802.01530

[Vor99]   A.A. Voronov, The Swiss-cheese operad, in *Homotopy Invariant Algebraic Structures*
            (Baltimore, 1998). Contemporary Mathematics, vol. 239 (Americal Mathematical
            Society, Providence, 1999), pp. 365–373. https://doi.org/10.1090/conm/239/03610.
            arXiv: math/9807037

[Whi41]   J.H.C. Whitehead, On adding relations to homotopy groups. Ann. Math. 2nd ser. **42**,
            409–428 (1941). ISSN: 0003-486X. https://doi.org/10.2307/1968907

[Whi49]   J.H.C. Whitehead, Combinatorial homotopy. I. Bull. Amer. Math. Soc. **55**, 213–245
            (1949). ISSN: 0002-9904. https://doi.org/10.1090/S0002-9904-1949-09175-9

[Wil14]   T. Willwacher, M. Kontsevich's graph complex and the Grothendieck–Teichmüller
            Lie algebra. Invent. Math. **200**(3), 671–760 (2014). ISSN: 1432-1297. https://doi.org/
            10.1007/s00222-014-0528-x. arXiv: 1009.1654

[Wil15]   T. Willwacher, Models for the *n*-Swiss Cheese operads (2015). arXiv: 1506.07021.
            Pre-published

[Wil16]   T. Willwacher, The homotopy braces formality morphism. Duke Math. J. **165**(10),
            1815–1964 (2016). https://doi.org/10.1215/00127094-3450644. arXiv: 1109.3520

[Wil17]   T. Willwacher, (Non-)formality of the extended Swiss Cheese operads (2017). arXiv:
            1706.02945. Pre-published

[Wil18]   T. Willwacher, Little disks operads and Feynman diagrams, in *Proceedings of the
            International Congress of Mathematicians* (Rio de Janeiro, 2018), vol. 2 (2018),
            pp. 1259–1280. https://www.mathunion.org/icm/proceedings

[Živ20]   M. Živković, Multi-directed graph complexes and quasi-isomorphisms between them
            I. Oriented graphs. High. Struct. **4**(1), 266–283 (2020). arXiv: 1703.09605

[Živ21]   M. Živković, Multi-directed graph complexes and quasi-isomorphisms between them
            II. Sourced graphs. Int. Math. Res. Not. **2**, 948–1004 (2021). ISSN: 1073-7928.
            https://doi.org/10.1093/imrn/rnz212. arXiv: 1712.01203

Index

© The Author(s), under exclusive license to Springer Nature Switzerland AG 2022
N. Idrissi, *Real Homotopy of Configuration Spaces*, Lecture Notes
in Mathematics 2303, https://doi.org/10.1007/978-3-031-04428-1

forwarded to the LNM Editorial Board, this is very helpful. If no reports are forwarded or if other questions remain unclear in respect of homogeneity etc, the series editors may wish to consult external referees for an overall evaluation of the volume.

5. Manuscripts should in general be submitted in English. Final manuscripts should contain at least 100 pages of mathematical text and should always include

   – a table of contents;
   – an informative introduction, with adequate motivation and perhaps some historical remarks: it should be accessible to a reader not intimately familiar with the topic treated;
   – a subject index: as a rule this is genuinely helpful for the reader.
   – For evaluation purposes, manuscripts should be submitted as pdf files.

6. Careful preparation of the manuscripts will help keep production time short besides ensuring satisfactory appearance of the finished book in print and online. After acceptance of the manuscript authors will be asked to prepare the final LaTeX source files (see LaTeX templates online: https://www.springer.com/gb/authors-editors/book-authors-editors/manuscriptpreparation/5636) plus the corresponding pdf- or zipped ps-file. The LaTeX source files are essential for producing the full-text online version of the book, see http://link.springer.com/bookseries/304 for the existing online volumes of LNM). The technical production of a Lecture Notes volume takes approximately 12 weeks. Additional instructions, if necessary, are available on request from lnm@springer.com.

7. Authors receive a total of 30 free copies of their volume and free access to their book on SpringerLink, but no royalties. They are entitled to a discount of 33.3 % on the price of Springer books purchased for their personal use, if ordering directly from Springer.

8. Commitment to publish is made by a *Publishing Agreement*; contributing authors of multiauthor books are requested to sign a *Consent to Publish form*. Springer-Verlag registers the copyright for each volume. Authors are free to reuse material contained in their LNM volumes in later publications: a brief written (or e-mail) request for formal permission is sufficient.

Addresses:
Professor Jean-Michel Morel, CMLA, École Normale Supérieure de Cachan, France
E-mail: moreljeanmichel@gmail.com

Professor Bernard Teissier, Equipe Géométrie et Dynamique,
Institut de Mathématiques de Jussieu – Paris Rive Gauche, Paris, France
E-mail: bernard.teissier@imj-prg.fr

Springer: Ute McCrory, Mathematics, Heidelberg, Germany,
E-mail: lnm@springer.com

Printed in the United States
by Baker & Taylor Publisher Services